Springer Theses

Recognizing Outstanding Ph.D. Research

Aims and Scope

The series "Springer Theses" brings together a selection of the very best Ph.D. theses from around the world and across the physical sciences. Nominated and endorsed by two recognized specialists, each published volume has been selected for its scientific excellence and the high impact of its contents for the pertinent field of research. For greater accessibility to non-specialists, the published versions include an extended introduction, as well as a foreword by the student's supervisor explaining the special relevance of the work for the field. As a whole, the series will provide a valuable resource both for newcomers to the research fields described, and for other scientists seeking detailed background information on special questions. Finally, it provides an accredited documentation of the valuable contributions made by today's younger generation of scientists.

Theses may be nominated for publication in this series by heads of department at internationally leading universities or institutes and should fulfill all of the following criteria

- They must be written in good English.
- The topic should fall within the confines of Chemistry, Physics, Earth Sciences, Engineering and related interdisciplinary fields such as Materials, Nanoscience, Chemical Engineering, Complex Systems and Biophysics.
- The work reported in the thesis must represent a significant scientific advance.
- If the thesis includes previously published material, permission to reproduce this must be gained from the respective copyright holder (a maximum 30% of the thesis should be a verbatim reproduction from the author's previous publications).
- They must have been examined and passed during the 12 months prior to nomination.
- Each thesis should include a foreword by the supervisor outlining the significance of its content.
- The theses should have a clearly defined structure including an introduction accessible to new PhD students and scientists not expert in the relevant field.

Indexed by zbMATH.

Daniele Cuomo

Architectures and Circuits for Distributed Quantum Computing

Doctoral Thesis accepted by
University of Parma, Parma, Italy

Springer

Author
Dr. Daniele Cuomo
Department of Computing Engineering
Valencia Polytechnic University
Valencia, Spain

Supervisor
Prof. Michele Amoretti
Department of Engineering
and Architecture
University of Parma
Parma, Italy

ISSN 2190-5053 ISSN 2190-5061 (electronic)
Springer Theses
ISBN 978-3-031-73807-4 ISBN 978-3-031-73808-1 (eBook)
https://doi.org/10.1007/978-3-031-73808-1

This Springer imprint is published by the registered company Springer Nature Switzerland AG
The registered company address is: Gewerbestrasse 11, 6330 Cham, Switzerland

Foreword

Distributed quantum computing (DQC) concerns with the execution of quantum algorithms over networked quantum computers. Using several small, high-quality quantum nodes instead of a single large quantum node has several advantages, including cost, scalability and correctness of the computation. There are also several challenges, as compiling and scheduling become more difficult when the network must be considered.

Daniele Cuomo's thesis opens interesting new perspectives on the design and implementation of DQC systems, using powerful tools like group theory and ZX-calculus. The models that are proposed shed some light on the inter-dependence between quantum computation and communication.

Chapter 1 presents some technologies and components for distributed quantum computing. The need for full system co-design is clearly illustrated and motivated. Furthermore, a logical network design tool is introduced, which is very useful for encoding network architectures and analyse their dynamics. I absolutely appreciated such a comprehensive view of DQC architectures.

Chapter 2 deals with the quantum circuit model of computation, ranging from fundamental concepts like reversibility and universality to more advanced ones like global gates and transpilation. An appendix provides a very nicely presented overview on the indefinite causal order framework, concerning quantum evolutions in which two or more operators occur, in an order defined by an extra quantum system.

Chapter 3 focuses on entanglement-based computation, which is the core of distributed quantum computing. Teleportation and TeleGate, which are major tools for implementing non-local gates, are nicely introduced. Entanglement paths are introduced, showing that their depth is constant independently of the number of middle processors. Furthermore, an efficient method for distributing global gates over multiple processors is introduced. Finally, groups of circuits that are relevant in the context of universal computation are identified, preparing the ground for a rigorous compilation model that will be presented in the last part of the thesis.

Chapter 4 is all about quantum noise, summarizing fundamental concepts that concern faulty gate modelling and error correction. In the appendix, a contribution to noise cancelling through the indefinite causal order framework is proposed. Here,

Daniele Cuomo shows evidence of deep knowledge of the problem and of the most relevant techniques for quantum error correction.

Finally, Chap. 5 contains the major contribution of the thesis. A novel formulation of the compilation problem as an optimization problem is proposed. Then, clever techniques for finding the best solutions (i.e. those with increased parallelism among quantum gates) are described. The proposed compiler is evaluated considering both Clifford and universal circuits generated at random.

I do not want to anticipate all the nice results proposed in this thesis. I am pretty sure that both advanced students and researchers in quantum computing will appreciate the hard work done by Daniele Cuomo.

Parma, Italy
August 2024

Prof. Michele Amoretti

Abstract

Quantum computers based on distributed architectures promise to be more *scalable* and less *fault-prone* than single-core quantum computers.

In accordance with novel technologies under development worldwide, *telegates* represent the fundamental operations supplied by distributed systems.

We give a comprehensive overview of distributed quantum computers based on telegates.

With some of the best tools for reasoning—i.e. network optimization, circuit manipulation, group theory and **ZX**-calculus—we have found new perspectives on the way a distributed quantum computer should be developed.

Contents

Chapter 1
Introduction

Distributed quantum computing stands as a pivotal applications in the field of quantum technologies. Distributed architectures could serve as our gateway to transcend the current NISQ era [1–6]. The integration of spatially distributed quantum processors presents an opportunity to construct scalable architectures that can effectively address the inherent challenges of quantum noise [4, 7]. However, the endeavor of interconnecting these distributed quantum processors is confronted with a set of significant challenges [2, 8–12].

To establish a practical distributed quantum computing system, it is imperative to stay abreast of state-of-the-art technologies under development worldwide. As such, this work commences by reviewing some cutting-edge hardware technologies—Sect. 1.1. This provides us (and the reader) with a realistic understanding of the fundamental components that comprise a distributed quantum computing system. Following this, we propose a modular and adaptable full-stack development in Sect. 1.2. In fact, This framework builds upon prior propositions [2].

With Chaps. 2 and 3, we shift our focus from hardware technologies to the fundamental tools that underpin distributed computing paradigms. We employ the widely accepted standard quantum circuit model to articulate quantum computation, beginning with the basics of unitary synthesis and decomposition. These fundamentals are not only crucial for local computations but also form the basis for understanding distributed architectures.

For the sake of completeness, Chap. 4 provides a framework for addressing quantum noise, arguably one of the most formidable challenges to date, given its inherent scalability issues with hardware. This chapter offers a perspective on the magnitude of the noise problem, which will inevitably be a component of any distributed computing system. The framework is bolstered by experimental results.

Drawing upon the knowledge gained throughout the preceding chapters, we conclude with numerical modeling and evaluation for complex scenarios of general

D. Cuomo, *Architectures and Circuits for Distributed Quantum Computing*,
Springer Theses, https://doi.org/10.1007/978-3-031-73808-1_1

interest—Chap. 5. We employ some of the most advanced tools for logical reasoning, such as network optimization, circuit manipulation, group theory, and the ZX-calculus. We meticulously identified and addressed numerous challenges, gaining valuable insights on how a distributed quantum computing system should be developed.

1.1 Technologies for Distributed Quantum Computing

1.1.1 Stationary-Flying Transduction

Qubit-qubit interaction generally works by means of some *transducer*. A transducer can be seen as a physical interface "converting quantum signals from one form of energy to another" [13]. It is especially true, in a distributed setting, that a transducer is able to move an information stored into some *stationary qubit*—e.g. a trapped-ion, a transmon or a quantum dot—into some flying object, usually photons. A photon is therefore an information carrier or medium, able to cover a long distance. Therefore, the medium can be used to make distant qubits interact.

The ability to engineer efficient transducers allows us to rethink at quantum architectures as to be scalable and modular. Depending on the transducer [13–22], different kinds of distributed architecture arise. For the sake of understanding qubit-qubit interaction in a distributed setting, we now consider distributed ion-traps architectures.

Scaling up a single ion-trap is challenging [23]. On the other hand, they represent a promising technology for integration within a distributed architecture, as a result of high gate fidelity [24, 25] and long life-time [26, 27]. In what comes next, we consider a cavity-based integration.

1.1.2 Cavity-Optical Coupling

Considering the scenario of qubits stored on different processors, to couple them, the physical setting needs to *scatter* quantum information outside a processor and reach the other one. This can be done by means of a single photon, canalized within an optical fiber.

In order to achieve such a configuration, we here consider ions able to be modeled as a three-levels system—see Fig. 1.1. Such a system depicts the experimental set-up proposed in [28].[1] The specifics of the system comes from the *ion species* selected to encode quantum information [31]. By placing such an ion within a *cavity*, this creates an ion-cavity system, where now the ion interact with the cavity mode. The

[1] The interested reader can find other settings at Refs. [29, 30].

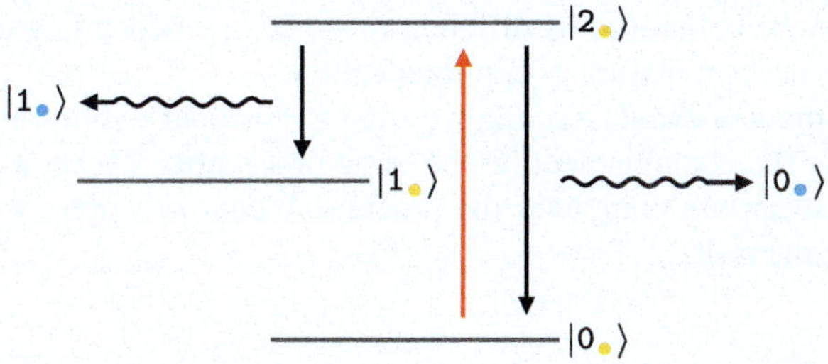

Fig. 1.1 Simplified energy level structure of an ion ($\bullet$) and relative photon emission ($\bullet$)

Fig. 1.2 Exemplary representation of an ion-cavity system emitting a photon. For technical details the reader may start from [32]

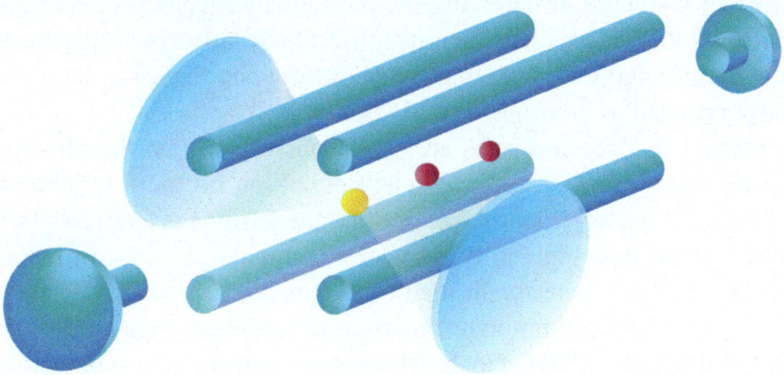

Fig. 1.3 An ion-trap embedded with a cavity pointing at a single ion. Representation inspired by a linear design [31, 35]

cavity has the role of collecting and scatter outside the system the photon emitted by the ion. Figure 1.2 shows a pictorial representation of the ion-cavity able to do so.

The first step taking place is the excitation of the ion $|0_{\bullet}\rangle \rightarrow |2_{\bullet}\rangle$—i.e. the red arrow in Fig. 1.1. Ideally, a spontaneous decay of the ion brings its energy with equal probabilities to one of the two lowest (and computationally relevant) states—i.e. $|0\rangle_{\bullet}$ and $|1_{\bullet}\rangle$. Furthermore, this happens with the emission of a photon which is coherent with the state of the ion.

As we anticipated, the scattered photon can be canalized within an optical fiber; the final configuration of the ion-fiber system is in the superposition $^1/_{\sqrt{2}}(|0_{\bullet}0_{\bullet}\rangle + |1_{\bullet}1_{\bullet}\rangle)$.

A pictorial representation of a single node of the distributed architecture is shown in Fig. 1.3. The cavity is pointing at one of the ions, which is coloured differently as an ion-trap may be composed by different ion species,[2] depending on whether it is meant to perform computation or communication.

To achieve the *non-local coupling* we need to consider two ion-traps generating and distributing the entanglement (at the same time), after which, a protocol called *entanglement swapping* completes the process. A *control system* will take care of accomplishing the task.

1.1.3 Control System

Summing up, to establish entanglement, the ions are simultaneously excited to an electronic state that spontaneously decays, during which a single photon each is emitted whose polarization is entangled with the ion's internal state. These photons are collected into optical fibres using free-space optics and sent to a common *terminal*.

The terminal take care of detecting the photons by means of a probabilistic *Bell state measurement*. This projects the ions into a maximally entangled state, heralded by the coincident detection of the pair of photons—see Sect. 1.1.4 for details. This is commonly referred as *entanglement swapping*.

It is important that the ion-traps are synchronized, so that the photon reach the terminal at the same time.[3] Classical synchronization protocol would take care of this by means of a master-clock. An experimental settings is available in Ref. [37].

Ultimately, to achieve scalability we need to consider the case of several processors. In the most basic scenario, all the processors are *centralized*, in the sense that all of them are wired to a common terminal to perform bell state measurement. Such a setting is the first example of scalable distributed architectures. The problems arising from such a setting is a *scheduling* problem. A multiplexer taking deterministic choices would be enough to ensure that all the processors are carefully scheduled to not create overlaps. An experimental settings, where four processors are scheduled by means of a multiplexer, is available in Ref. [38].

1.1.4 Bell State Analyser

To keep the discussion easy we explained the main procedures by means of three-levels systems for the ions. However, each of this state should be split to create several possible configurations. This doesn't change the whole protocol, but it has consequences on the possible outcomes obtained by measuring the photons. Different

---

[2] Which brings to the classification in *communication* and *computation* (or data) qubits [2, 5, 33, 34].

[3] Without synchronization, one can retrieve the likelihood of each photon source. This makes the photons distinguishable, causing a loss in fidelity [36].

ion species and kind of measurement lead to different configurations. In general, we distinguish three main outcomes:

- One or both photons failed to be detected by the measurement. This means that the whole procedure is basically wasted time, as the nodes need to attempt again from scratch. This possibility can be a serious problem to practical computation, as a low *success rate* leads to long waiting times.
- The protocol succeeded and the ions are in one of the four Bell states $\{|\Phi^+\rangle, |\Phi^-\rangle, |\Psi^+\rangle, |\Psi^-\rangle\}$.
- The protocol partially succeeded. Namely, a superposition between two bell states has been created, e.g. $|\Phi^{\pm}\rangle$ or $|\Psi^{\pm}\rangle$. This scenario may occur for several reasons—depending also from the employed physical settings—it can be caused by *dark measurements* [36], a rare and negligible scenario. Otherwise, it may come, for example, from two *clicks* coming from the same detector.

Different Bell measurement settings brings to different sets of heralded entanglements [39–42]. We here consider the quite general case of Ref. [39], as we think this case may be particularly efficient for distributed computation. A pictorial representation of the setting is reported in Fig. 1.4. In fact such a configuration brings to a 50% chance of success—i.e. two clicks on different detectors—and 50% of partial success—i.e. two clicks on the same detector. Namely, when the protocol succeeds, the final state is in $\{|\Psi^-\rangle, |\Psi^+\rangle\}$, while in case of partial success, the output has an ambiguous phase: $|\Phi^{\pm}\rangle$.

The ideal result coming from performing entanglement generation and distribution followed by entanglement swapping is a maximally entangled state between distant qubits. In practice, this is not achievable as each of the complicated techniques we described are in general not perfect, resulting in a state slightly different from a Bell pair. One can evaluate the final distributed state in terms of *fidelity* with some target Bell state. E.g.,

$$\mathfrak{f} = \langle\Phi^+|\,\sigma\,|\Phi^+\rangle\,. \tag{1.1}$$

where σ is the generated state. For example, consider the experiments reported in Ref. [39, 40]. The author's proposal starts with the generation of a non-maximally entangled state ion-photon

$$\sqrt{\frac{2}{3}}\,|0_{\bullet}0_{\bullet}\rangle + \sqrt{\frac{1}{3}}\,|1_{\bullet}1_{\bullet}\rangle\,. \tag{1.2}$$

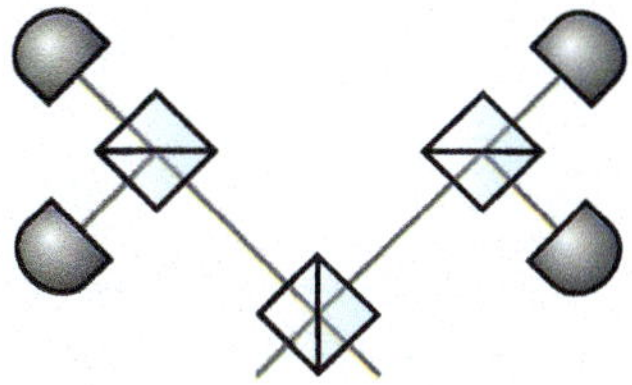

Fig. 1.4 A common setting for the Bell state measurement [39]

However, after the collection into the single mode fiber, the state gets projected to the Bell pair

$$\frac{1}{\sqrt{2}}(|0_{\circ}0_{\bullet}\rangle + |1_{\circ}1_{\bullet}\rangle). \tag{1.3}$$

Once performed the entanglement swap, the ion-ion average fidelity is 0.94; a promising result. Unfortunately, in the perspective of practical computation, the fidelity needs to be some $\mathfrak{f} = 1 - \varepsilon$ with ε small enough to keep the error rate *manageable*, e.g. by means of *error correction schemes*—treated in Chap. 4. A possible solution is called *entanglement distillation* [43–45] (or purification). However, choosing the best approach is not trivial and may very depend on the architecture specifics.

1.2 Full System Co-Design

We reported several important research fields deeply related to the implementation of a first distributed and scalable architecture. We introduced the required technologies to achieve qubit-qubit interaction when these are arbitrarily far apart.

The considered literature gives a perspective on how to integrate multiple quantum processors into a scalable architecture, able to perform distributed quantum computation.

Stemming from the above overview, we now need to extract a full-stack development, by identifying the most important roles—and dependencies—and we will need to engineer an *ecosystem* [2], providing a framework for distributed quantum computation. As we are facing the early stage of quantum computation and distributed architectures, it is wise to focus on the main challenges and assigning them to a proper *entity*.[4] In fact, any proposal now is highly prone to changes, because of the continuous growing of the field [3] and the huge advancing in both technology and information theory.

We propose a design that starts from a bottom-up reasoning, stacking up a number of layers where the lower ones provide some *resources* and flexibility which the upper layers can rely on. More precisely, consider Fig. 1.5: this shows a linear stack where each layer represents one of the fundamental subjects necessary to create a practical framework.

---

[4] As usual in the engineer terminology, an entity is something quite abstract, which needs to be defined in terms of its roles and relations with other entities.

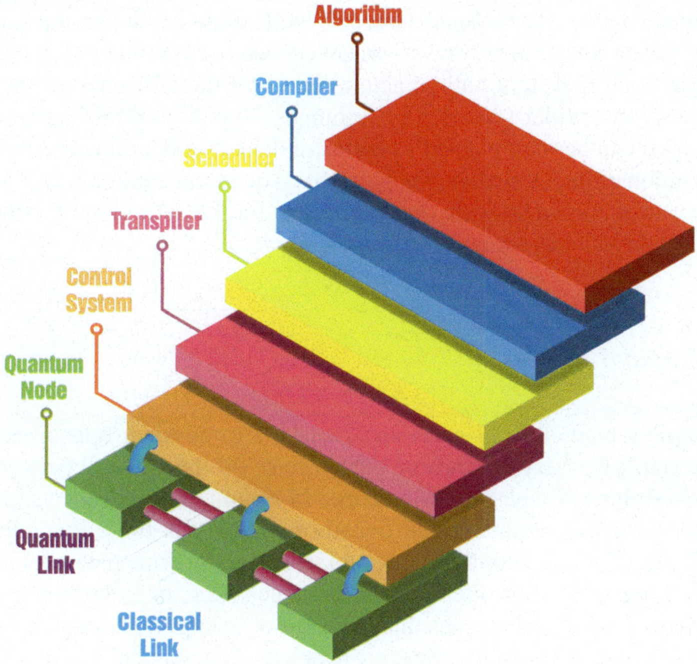

Fig. 1.5 Full-stack development of a distributed quantum computing framework

1.2.1 Physical Layers

As usual, in architecture design, the first layer represents the hardware. For the sake of clarity, we depicted a network composed of three quantum nodes. We already discussed how such a network may be achieved by focusing on ion-traps integrated with cavities, Bell state analysers and multiplexers—see Sects. 1.1 and 1.1.4.

A quantum node refers to the full hardware set-up working locally, which includes, of course, the quantum processor. The nodes are inter-connected by **quantum links** wherein mediums carries quantum information. Such a set-up can be achieved, for example, by means of [38]:

- ion-traps, cavity-based transducers as quantum nodes and
- optical fibers, multiplexers and bell-state analysers as quantum links.

As explained in Sect. 1.1.3, as minimum requirement for the system to be *operative*, the network needs to be carefully handled by a classical **control system**, which deals with real-time protocol synchronization. Because of its role, the control system is mainly physical. In fact, it is directly connected to each quantum node by means of **classical links**. The linkage will be also used to gather classical information coming from *measurement-based computation* [46].

Advancements in the physical network[5] will allow to *evolve* the layer, up to becoming a *quantum control system*—as envisioned for example in Refs. [50–53]—, a quantum control system will be able to optimize the efficiency of the network, having access to a wider spectrum of resources—e.g. high-dimensional entangled states—which can be manipulated to optimize the network efficiency, by means of communication protocol fundamentally based on quantum communication theory. A quantum control system may also be responsible for the definition of a (distributed) error correction scheme—treated in Chap. 4.

1.2.2 Logical Layers

The **transpiler** maps logical gates and protocols to physical operations. For this reason the transpiler very depends on the specifics for the physical network.

The **scheduler** we envision is in charge of optimizing network usage, also by taking advantage of possible circuit transformation at run-time. Regardless of the specific design for the scheduler, this would inherently provide the upper layers with a simplified perspective for the network dynamics—e.g. as to be functioning over *discrete time domain*. This means that any kind of unexpected delay or failure gets negligible to the upper-layers, as these are handled by the scheduler and/or the control system.

The **compiler**'s task is to map an *hardware-agnostic algorithm* to some equivalent form, which is better suited for distributed computation.

The upper layer is an **algorithm**, an abstract input, which the framework takes charge of and carefully spread throughout the whole stack in order to be processed.

Once the framework is ready-to-go, it will be able to accept some *groups of algorithms*,[6] up to *universal* groups. Refer to Chap. 2 for details.

1.3 Network Encoding

A **logical network** is a simplified representation of the physical architecture, which is synthesised to a few main features. Namely, (i) a topology, describing the possible interactions among nodes, and (ii) the kinds of interaction, referred as *logical gates*. The set of logical gates will be the fundamental components upon which building a *computational paradigm*.

Logical gates can be local—e.g. multi-qubit gates [55–58]—and non-local—i.e. telegates—. In fact, while local gates operate on logical nodes, telegates operates throughout the logical network. This makes the latter resource much more expensive

[5] Up to the development of a *quantum internet* [2, 5, 8, 10, 11, 47–49].

[6] Parallel-based algorithms—e.g. see Ref. [54] for a parallel algorithm solving the quantum Fourier transform—are natively meant to work on distributed architectures.

and worth of dedicated analysis; especially considering the degree of novelty in the context of computational paradigms that may arise. Refer to Sect. 3 for details.

In accordance with the current trend on quantum technologies, above reported, we now propose a mathematical description of the logical network. Formally, let $\mathcal{N} = (V, P, F)$ be a network triple representing the architecture. $V = Q \cup C$ is a set of nodes describing qubits, therefore it is the disjoint union of computation qubits $Q = \{q_1, q_2, \ldots, q_{|Q|}\}$ and communication qubits $C = \{c_1, c_2, \ldots, c_{|C|}\}$. We can represent n processors by partitioning V into $P = \{P_1, P_2, \ldots, P_n\}$. Therefore, a sub-set P_i characterizes a processor as its set of qubits/nodes.

$F = L \cup R$ is as a set of undirected edges. L represents the local couplings, therefore

$$L \subseteq \bigcup_i P_i \times P_i.$$

Notice that there is no particular assumption on connectivity nor cardinality within processors. This keeps the treating hardware-independent and it allows for heterogeneous architectures.

R represents entanglement links. Since entanglement links connect only communication qubits, we introduce, for each processor, a set of those qubits only; i.e., $C_i = C \cap P_i$. Therefore, we have

$$R \subseteq \bigcup_{i,j\,:\,i \neq j} C_i \times C_j.$$

Figure 1.6 shows an exemplary architecture, with three processors in P, six computation qubits in Q, six communication qubits in C, three entanglement links in R and ten local couplings in L.

Concerning minimal assumptions, we only care about architectures actually able to perform any operation. This translated into a simple inter-processor connection assumption. As regards local coupling, our work covers mainly (but is not limited to) fully connected processors. Ion-traps already satisfy such an assumption [59],

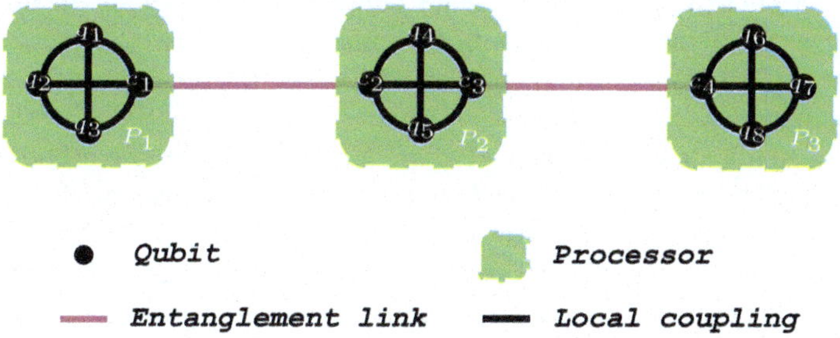

Fig. 1.6 Toy distributed quantum architecture with 3 processors

but there may be others in the future to join. Consider as an example that, for superconductor processors, it has already been developed a high connectivity topology, which has been proposed together with a blueprint to achieve full connectivity [60].

In order to keep consistency between the physical architecture and its logical representation, it is mandatory to identify an array of essential tasks and design a full system, that orchestrate these tasks, providing at last, a framework for distributed quantum computation.

1.3.1 Time Domains and Network Dynamics

When the computation is running, the system we described starts working, which inherently brings us to give a definition of time. A time domain may depend on whether this describes a logical network or a physical one—such as the control system.

It should be clear that the entanglement link generation has central interest in our treating. In fact, we are also going to scan the time as such links go established. Specifically, notice that link generations among different couples of qubits are independent. For this reason we assume that all the possible links get generated simultaneously and, as soon as all the states are measured, a new round of simultaneous generations begins. We define a primitive for the generic link generation, which we will use throughout this thesis:

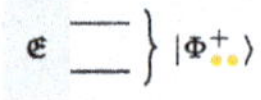

For the sake of reasoning, the logical network dynamics is scanned by the parallel generations of such a primitive. Namely, a unit of time is the lapse the system need to establish one link. Any link that the the physical network is able to provide is, in principle, available. From these some are generated in parallel, consumed and then the network is ready to re-generate other links in the subsequent unit of time.

It is important to notice that there is not a unique way of defining the time domain, and taking a choice may bring to critical limitations for the overall performance of the system. It is not immediate to understand the specific roles that each layer should be responsible of.

Given the early stage of distributed quantum technologies, we propose to model the logical dynamics to be agnostic about potential faults occurring during link generation. A unit of time *roughly* corresponds to the *heralding rate*. With this choice, we are overloading the physical layers, which are tasked with providing a fault-robust network. While this assumption may seem like an oversimplification, for now, it helps us identify what are the most important features characterizing distributed computation. Nevertheless, in the future, a more advanced design would render logical layers resilient to faults.

1.3.2 Architecture Encoding for Design Evaluation

Once every element comprising the full system is introduced, it is useful to concentrate on specific configurations that offer us a behavioral model representing distributed quantum computation. Specifically, we now provide the specifics for the logical network we will use to evaluate our design.

A generic quantum processor is, as before, made of computation and communication qubits, with an all-to-all local connectivity. We assume that a quantum node has a single computation qubit. This assumption is reasonable as it stresses the architecture to be meant for distributed computation. Also, any extra qubit wouldn't be wasted, as fault-tolerance is very resource-demanding. Hence, a quantum processor would benefit from the extra qubits as this are devoted to enhance the fidelity of computation.

As regard the inter-processor connectivity, downstream of the topologies evaluation made in [33], in this thesis we will consider networks that can be described as *grid lattices*.[7] We assume that each processor has enough communication qubits to guarantee the logical network's structure.

In light of the above observations, it is reasonable and convenient to consider the whole processor as a *node*, and define a function c that provides the number of available links between two processors. Specifically, we first formalized a distributed architecture as the network graph $\mathcal{N} = (V, P, F)$ introduced in Sect. 1.3; this step was important to understand the interior behavior of remote operations from a qubit perspective. However, it is useful to re-state it to a more compact encoding. Formally speaking, we will consider a *quotient graph* of $\mathcal{N}$.

To not further weigh down the formalism, we re-model a logical network, by considering the processors as nodes, corresponding to an enumeration for the partition P, i.e., $\mathfrak{p} = \{\mathfrak{p}_1, \mathfrak{p}_2, \ldots, \mathfrak{p}_n\}$. All the entanglement links, connecting the same couple of processors, now collapse to an only edge with integer capacity c, describing how many parallel entanglement links the two processors supplies. We refer to this sets of edges as

$$\mathfrak{l} \subseteq \bigcup_{i,j\ :\ i \neq j} \mathfrak{p}_i \times \mathfrak{p}_j.$$

Hence, the new undirected graph is $\mathcal{Q} = (\mathfrak{p}, \mathfrak{l})$.

In Fig. 1.7 we show the quotient graph related to the toy architecture of Fig. 1.6. Throughout this thesis we will focus on edges of capacity $c = 1$, omitting so the capacity function from the network definition.

Putting things together, the control system provides a **discrete** perspective of the **dynamic** network. At the beginning of a time-step $\mathfrak{t}$, the logical network corresponds to a grid of available entanglement links, such as the small grid shown in Fig. 1.8. As the computation runs, at each time-step a subset of $\mathcal{N}$ is selected to perform inter-processors operations. See for example Fig. 1.9.

---

[7] As reported in [33], a grid connectivity is convenient both for physical implementation and logical optimization.

Fig. 1.7 Quotient graph derived from Fig. 1.6. The processors become the nodes, the entanglement links between a couple of processors gather into one edge, with capacity equal to the number of original links [33]

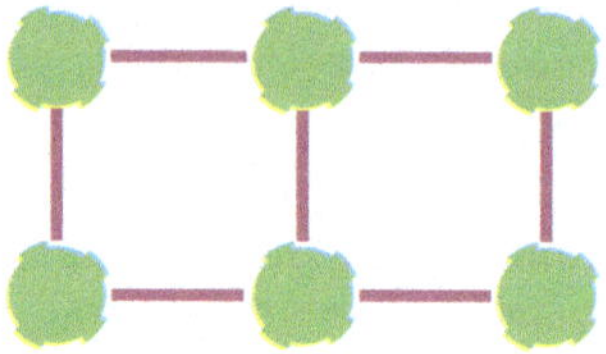

Fig. 1.8 Exemplary grid lattice at any given time-step t

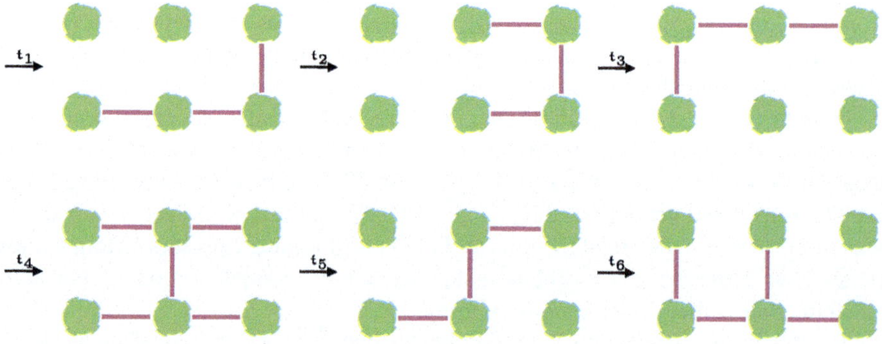

Fig. 1.9 Dynamics of a grid-like network. This exemplary computation lasts 6 time-steps and, at each time-step, some edges get selected to perform inter-processor operations

The reader can notice that, when selecting edges, the computation gets characterized by *paths* and *trees*. Within next chapter, we will inspect the protocols—i.e. *entanglement path* and *entanglement tree*—that ensure such data structures are consistent representations of inter-processor operations.

References

1. Preskill J (2018) Quantum computing in the NISQ era and beyond. Quantum 2(79)
2. Cuomo D, Caleffi M, Cacciapuoti AS (2020) Towards a distributed quantum computing ecosystem. IET Quantum Commun 1(1):3–8
3. Gibney E (2019) The quantum gold rush. Nature 574(7776):22–24
4. Gambetta J (2022) Expanding the IBM Quantum roadmap to anticipate the future of quantum-centric supercomputing
5. Cacciapuoti AS et al (2019) Quantum internet: networking challenges in distributed quantum computing. IEEE Netw 34(1):137–143

6. Van Meter R, Devitt SJ (2016) The path to scalable distributed quantum computing. Computer 49(9):31–42
7. Eddins A et al (2022) Doubling the size of quantum simulators by entanglement forging. PRX Quantum 3(1):010309
8. Jeff Kimble H (2008) The quantum internet. Nature 453(7198):1023–1030
9. Pirandola S, Braunstein SL (2016) Physics: Unite to build a quantum Internet. Nat News 532(7598):169
10. Dür W, Lamprecht R, Heusler S (2017) Towards a quantum internet. Europ J Phys 38(4):043001
11. Wehner S, Elkouss D, Hanson R (2018) Quantum internet: a vision for the road ahead. Science 362(6412)
12. Castelvecchi D (2018) The quantum internet has arrived (and it hasn't). Nature 554(7690):289–293
13. Lauk N et al (2020) Perspectives on quantum transduction. Quantum Sci Technol 5(2):020501
14. Martin JA Schütz and Martin JA Schütz (2020) Universal quantum transducers based on surface acoustic waves. In: Quantum dots for quantum information processing: controlling and exploiting the quantum dot environment, pp 143–196
15. Brubaker BM et al (2022) Optomechanical ground-state cooling in a continuous and efficient electro-optic transducer. Phys Rev X 12(2):021062
16. McKenna TP et al (2020) Cryogenic microwave-to-optical conversion using a triply resonant lithium-niobate-on-sapphire transducer. Optica 7(12):1737–1745
17. Zeuthen E et al (2020) Figures of merit for quantum transducers. Quantum Sci Technol 5(3):034009
18. Kielpinski D, Monroe C, Wineland DJ (2002) Architecture for a large-scale ion-trap quantum computer. Nature 417(6890):709–711
19. Fang Y-L, Zheng H, Baranger HU (2014) One-dimensional waveguide coupled to multiple qubits: photon-photon correlations. EPJ Quantum Technol 1(1):1–13
20. Zhong C et al (2020) Proposal for heralded generation and detection of entangled microwave-optical-photon pairs. Phys Rev Lett 124(1):010511
21. Krastanov S et al (2021) Optically heralded entanglement of superconducting systems in quantum networks. Phys Rev Lett 127(4):040503
22. Gold A et al (2021) Entanglement across separate silicon dies in a modular superconducting qubit device. npj Quantum Inf 7(1):1–10
23. Moody G et al (2022) Roadmap on integrated quantum photonics. J Phys: Photon 4(1):012501
24. Harty TP et al (2014) blackHigh-fidelity preparation, gates, memory, and readout of a trapped-ion quantum bit. Phys Rev Lett 113(22):220501
25. Ballance CJ et al (2014) blackHigh-fidelity two-qubit quantum logic gates using trapped calcium-43 ions. arXiv preprint arXiv:1406.5473
26. Wehner S, Elkouss D, Hanson R (2018) Quantum internet: a vision for the road ahead. Science 362(6412):eaam9288
27. Wang P et al (2021) blackSingle ion qubit with estimated coherence time exceeding one hour. Nat Commun 12(1):1–8
28. Monroe C et al (2014) Large-scale modular quantum-computer architecture with atomic memory and photonic interconnects. Phys Rev A 89(2):022317
29. Gao S et al (2023) Optimization of scalable ion-cavity interfaces for quantum photonic networks. Phys Rev Appl 19(1):014033
30. Salmon W et al (2022) Gauge-independent emission spectra and quantum correlations in the ultrastrong coupling regime of open system cavity-QED. Nanophotonics 11(8):1573–1590
31. Bruzewicz CD et al (2019) Trapped-ion quantum computing: progress and challenges. Appl Phys Rev 6(2):021314
32. Kaya Ü. The laser optical cavity. https://chem.libretexts.org/@go/page/13651
33. Cuomo D et al (2023) Optimized compiler for distributed quantum computing. ACM Trans Quantum Comput (2023)
34. Caleffi M, Cacciapuoti AS (2020) Quantum switch for the quantum internet: noiseless communications through noisy channels. IEEE J Select Areas Commun 38(3):575–588

35. Paul W (1990) Electromagnetic traps for charged and neutral particles. Rev Modern Phys 62(3):531
36. Stephenson L (2019) Entanglement between nodes of a quantum network. PhD thesis. University of Oxford
37. Nadlinger DP et al (2022) Experimental quantum key distribution certified by Bell's theorem. Nature 607(7920):682–686
38. Oi DKL, Devitt SJ, Hollenberg LCL (2006) Scalable error correction in distributed ion trap computers. Phys Rev A 74(5):052313
39. Lütkenhaus N, Calsamiglia J, Suominen K-A (1999) Bell measurements for teleportation. Phys Rev A 59(5):3295
40. Stephenson LJ et al (2020) blackHigh-rate, high-fidelity entanglement of qubits across an elementary quantum network. Phys Rev Lett 124(11):110501
41. Valivarthi R et al (2014) Efficient Bell state analyzer for time-bin qubits with fast-recovery WSi superconducting single photon detectors. Opt Express 22(20):24497–24506
42. Mattle K et al (1996) Dense coding in experimental quantum communication. Phys Rev Lett 76(25):4656
43. Rozpedek F et al (2018) Optimizing practical entanglement distillation. Phys Rev A 97(6):062333
44. Kalb N et al (2017) blackEntanglement distillation between solid-state quantum network nodes. Science 356(6341):928–932
45. Hu X-M et al (2021) Long-distance entanglement purification for quantum communication. Phys Rev Lett 126(1):010503
46. Nielsen MA (2003) Quantum computation by measurement and quantum memory. Phys Lett A 308(2–3):96–100
47. Van Meter R et al (2022) A quantum internet architecture. In: 2022 IEEE international conference on quantum computing and engineering (QCE). IEEE, pp 341–352
48. Kozlowski W et al (2021) Architectural principles for a quantum internet. Internet-draft draft-irtf-qirg-principles-03. Work in progress. Internet engineering task force, June 2021, 39 pp
49. Pant M et al (2019) Routing entanglement in the quantum internet. npj Quantum Inf 5(1):1–9
50. Koudia S et al (2022) How deep the theory of quantum communications goes: superadditivity, superactivation and causal activation. In: IEEE communications surveys and tutorials
51. Illiano J et al (2022) Quantum internet: from medium access control to entanglement access control. arXiv preprint arXiv:2205.11923
52. Avis G, Rozpedek F, Wehner S (2022) Analysis of multipartite entanglement distribution using a central quantum-network node. arXiv preprint arXiv:2203.05517
53. Wallnöfer J et al (2019) (2019) Multipartite state generation in quantum networks with optimal scaling. Sci Rep 9(1):1–18
54. Cleve R, Watrous J (2000) Fast parallel circuits for the quantum Fourier transform. In: Proceedings 41st annual symposium on foundations of computer science. IEEE, pp 526–536
55. Lu Y et al (2019) Global entangling gates on arbitrary ion qubits. Nature 572(7769):363–367
56. Casanova J et al (2012) Quantum simulation of interacting fermion lattice models in trapped ions. Phys Rev Lett 108(19):190502
57. Ivanov SS, Ivanov PA, Vitanov NV (2015) Efficient construction of three-and four-qubit quantum gates by global entangling gates. Phys Rev A 91(3):032311
58. Martinez EA et al (2016) Compiling quantum algorithms for architectures with multi-qubit gates. New J Phys 18(6):063029
59. Brown KR, Kim J, Monroe C (2016) Co-designing a scalable quantum computer with trapped atomic ions. npj Quantum Inf 2(1):1–10
60. Hazra S et al (2021) Ring-resonator-based coupling architecture for enhanced connectivity in a superconducting multiqubit network. Phys Rev Appl 16(2):024018

Chapter 2
Computing with Quantum Circuits

2.1 Universality and Circuit Synthesis

In this section, we provide preliminary explanations that expound how a quantum processor can execute any algorithm using only a limited set of gates. Such a set is therefore called *universal*. We briefly get through the Boolean logic—which governs classical computation—and how we can express any Boolean function within the quantum framework. This gives provide the reader with a perspective of how quantum logic can express a wider group of functions.

2.1.1 Boolean Logic

Classical computation is deeply based on Boolean logic. Since classical technologies are really advanced and benefits from many years of research on physical implementations, several Boolean operators find direct implementation as classical gates. However, our interested is narrowed to how we can express any boolean function by means of quantum operators. Hence we can restrict the discussion to a single logical operator: $\neg(\mathtt{b}_1 \wedge \mathtt{b}_2)$. In fact, for any Boolean variables $\mathtt{b}_1, \mathtt{b}_2 \in \{\mathtt{0}, \mathtt{1}\}$, the $\neg(\mathtt{b}_1 \wedge \mathtt{b}_2)$ operator[1] is universal to Boolean logic [1], hence we need to find a quantum operator able to realize it. The Toffoli operator [2] does the job. Formally, a classical state $\mathtt{b}_1 \cdot \mathtt{b}_2$ gets encoded in the quantum state $|\mathtt{b}_1\rangle \otimes |\mathtt{b}_2\rangle \otimes |\mathtt{1}\rangle$. By applying the Toffoli operator to the encoded system, the last qubit encodes the Boolean state $\neg(\mathtt{b}_1 \wedge \mathtt{b}_2)$. This is shown in Fig. 2.1.

To date, there is no direct physical implementation for the Toffoli operator. It needs to be expressed as a composition of quantum operators physically realizable.

[1] The value is true iff no more than one variable is true.

D. Cuomo, *Architectures and Circuits for Distributed Quantum Computing*,
Springer Theses, https://doi.org/10.1007/978-3-031-73808-1_2

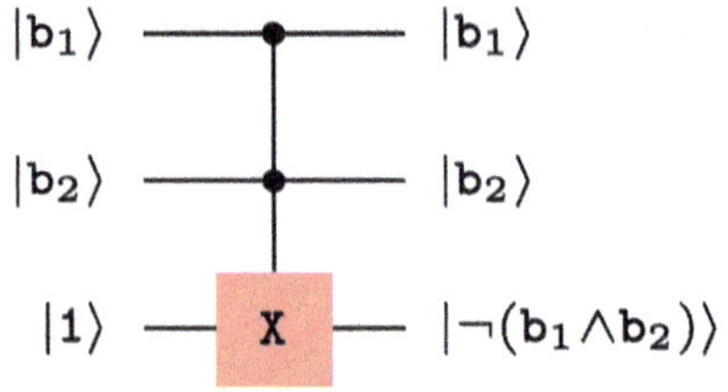

Fig. 2.1 $\neg(b_1 \wedge b_2)$ operator by means of a Toffoli

2.1.2 Quantum Logic

We here report a step-by-step introduction to quantum logic and what a quantum processors should have to provide universal computing. Quantum computing works with the logic of pure states. A Hilbert space $\mathbb{H}$ of dimension d is closed under the *unitary group* of degree d. This means that for any pure state $|\varphi\rangle \in \mathbb{H}$ and any unitary operator $\mathtt{U}$, it results $\mathtt{U}\,|\varphi\rangle \in \mathbb{H}$.

A generic quantum algorithm can be expressed as a system initialized to $|0\rangle^{\otimes d}$ and a unitary $\mathtt{U}$ operating on it. Figure 2.2 gives a circuit representation.

Since quantum processors do not supply as primitive operator a generic unitary, this is subject to one or more steps of *decomposition* or *synthesis*. A generic decomposition is showed in circuit of Fig. 2.3.

A universal operator set should be *efficient*, in the sense that the overhead caused by the decomposition from a d-degree unitary to the operator set is upper-bounded by some polynomial function. There are several (historically) important results showing how such a requirement is achievable. We start from one coming from the work done in [3, 4], showing that, given a generic 2-degree unitary $\mathtt{U}$—i.e. it operates over a single qubit –, the following operator is universal:

$$\wedge(\mathtt{U}) \equiv |0\rangle\langle 0| \otimes \mathbb{1} + |1\rangle\langle 1| \otimes \mathtt{U} \tag{2.1}$$

This is a controlled-$\mathtt{U}$ operator in Feynmann's notation [5]. We will use this notation throughout this thesis because of its versatility. Even if it is a bit outdated, we think

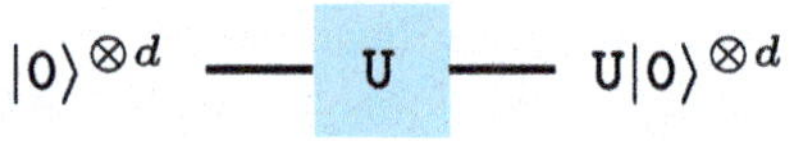

Fig. 2.2 Generic algorithm expressed as a unitary, operating over a $|0\rangle^{\otimes d}$ state

Fig. 2.3 Generic decomposition of $\mathtt{U}$ into k unitaries

Fig. 2.4 Circuit representation for the $\wedge(\mathtt{U})$ operator

it helps to highlight the logic behind the operators; it also helps us to keep consistency throughout the chapters. In the circuit model the subject operator appears as in Fig. 2.4.

One can notice that the group of equation (2.1) is pretty compact, as it involves only 2-qubit operators where one of them act always as control. Nevertheless, some decomposition step is necessary to get closer to what real processors can actually offer. To this aim, we need to introduce some further operators. The first one is known as *special unitary*. We refer to this operator as $\mathtt{V}_{\alpha,\beta,\delta}$ and it is defined as follows [6]:

$$\mathtt{V}_{\alpha,\beta,\delta} \equiv \begin{pmatrix} e^{\mathrm{i}(\alpha+\beta)/2} \cos {}^{\delta}/_{2} & e^{\mathrm{i}(\alpha-\beta)/2} \sin {}^{\delta}/_{2} \\ -e^{\mathrm{i}(-\alpha+\beta)/2} \sin {}^{\delta}/_{2} & e^{\mathrm{i}(-\alpha-\beta)/2} \cos {}^{\delta}/_{2} \end{pmatrix} \tag{2.2}$$

The second operator represent a *global phase shift*:

$$\mathtt{G}_{\gamma} \equiv \begin{pmatrix} e^{\mathrm{i}\gamma} & 0 \\ 0 & e^{\mathrm{i}\gamma} \end{pmatrix} \tag{2.3}$$

It results that special unitaries composed with global phase shifts characterizes the group of unitaries. It follows the same statement in matrix form:

$$\mathtt{U}_{\gamma,\alpha,\beta,\delta} \equiv \mathtt{G}_{\gamma}\mathtt{V}_{\alpha,\beta,\delta} \tag{2.4}$$

As consequence, the first decomposition we can apply comes from the circuit equivalence of Fig. 2.5.

A conditioned global phase shift is equivalent to a *relative phase shift*. Thus, let $\mathtt{R}_{\gamma}$ such an operator, defined as follows:

$$\mathtt{R}_{\gamma} \equiv \begin{pmatrix} 1 & 0 \\ 0 & e^{\mathrm{i}\gamma} \end{pmatrix} \tag{2.5}$$

Fig. 2.5 First decomposition as a result of equivalence (2.4)

Fig. 2.6 Circuit representation of the equivalence $\wedge(\mathrm{G}_\gamma) \equiv \mathrm{R}_\gamma \otimes \mathbb{1}$

The above statement allows us to *synthesize* $\wedge(\mathrm{G}_\gamma)$ into $\mathrm{R}_\gamma \otimes \mathbb{1}$. Figure 2.6 shows the corresponding circuit equivalence.

The final step to achieve universal computing through physically realizable operators is to decompose $\wedge(\mathrm{V}_{\alpha,\beta,\delta})$. To this aim, let us introduce the rotational operators over the Pauli axes. Namely, $\mathrm{X}_\gamma \equiv \mathrm{e}^{-\mathrm{i}\mathrm{X}\gamma/2}$, $\mathrm{Y}_\gamma \equiv \mathrm{e}^{-\mathrm{i}\mathrm{Y}\gamma/2}$ and $\mathrm{Z}_\gamma \equiv \mathrm{e}^{-\mathrm{i}\mathrm{Z}\gamma/2}$. Notice also that

$$\mathrm{e}^{-\mathrm{i}\mathrm{E}\gamma/2} = \cos \gamma/2 - \mathrm{i}\,\mathrm{E} \sin \gamma/2, \tag{2.6}$$

holds for any $\mathrm{E} \in \{\mathrm{X}, \mathrm{Y}, \mathrm{Z}\}$. Therefore, whenever $\gamma = \pi$, each rotational operator relates to the corresponding Pauli operator X, Y or Z. Formally, they are equivalent up to the global phase $\mathrm{G}_{-\pi/2} \equiv -\mathrm{i}$. E.g. $\mathrm{X}_\pi \cong \mathrm{X}$.

Now we proceed by reporting the results coming from [6]. Let $\mathrm{A}_{\alpha,\delta}$, $\mathrm{B}_{\delta,\alpha,\beta}$, $\mathrm{C}_{\alpha,\beta}$ be a triplet of special unitaries defined as follows:

- $\mathrm{A}_{\alpha,\delta} = \mathrm{Y}_{\delta/2}\mathrm{Z}_\alpha$;
- $\mathrm{B}_{\delta,\alpha,\beta} = \mathrm{Y}_{-\delta/2}\mathrm{Z}_{-(\alpha+\beta)/2}$;
- $\mathrm{C}_{\alpha,\beta} = \mathrm{Z}_{(\beta-\alpha)/2}$.

With such a triplet, together with $\wedge(\mathrm{X})$ we are able to decompose $\wedge(\mathrm{V}_{\alpha,\beta,\delta})$, according to the circuit equivalence shown in Fig. 2.7.

We can finally show the universality by composing the two results and getting a decomposition for $\wedge(\mathrm{U}_{\gamma,\alpha,\beta,\delta})$—see Fig. 2.8.

Theorem *The* IBM *gate set [7–11] is universal.*

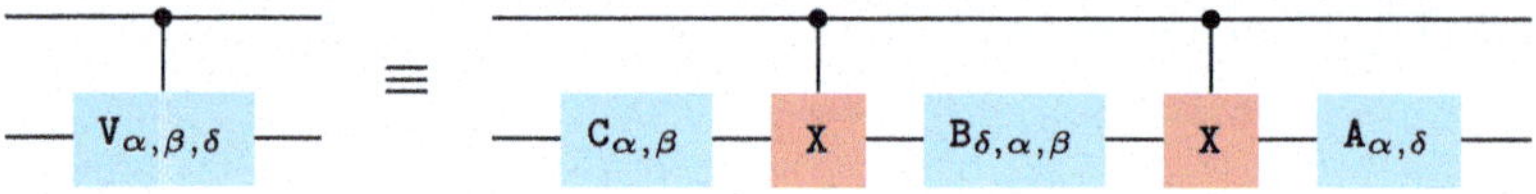

Fig. 2.7 Decomposition of $\wedge(\mathrm{V}_{\alpha,\beta,\delta})$

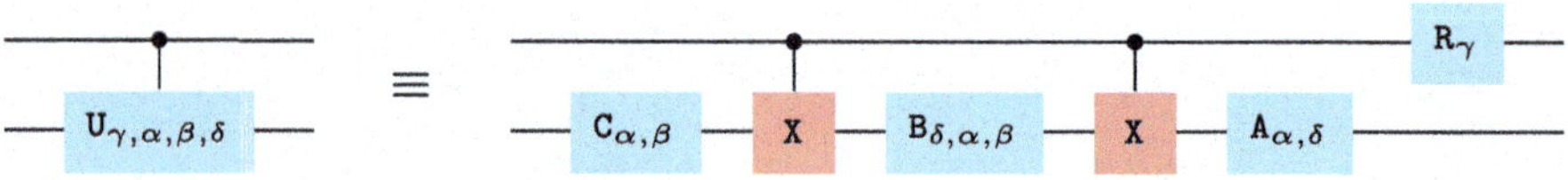

Fig. 2.8 Decomposition of $\wedge(\mathrm{U}_{\gamma,\alpha,\beta,\delta})$

Proof All the processors supplied by the IBM cloud have gate set $\{Z_\gamma, X_{\pi/2}, X, \wedge(X)\}$. To prove the universality of such a set, notice that $R_\gamma \cong Z_\gamma$. Furthermore, any special unitary $V_{\alpha,\beta,\delta}$ can be factorized as follows [6]:

$$V_{\alpha,\beta,\delta} \equiv Z_\alpha Y_\delta Z_\beta$$

Since an IBM processor can run natively a generic Z_γ gate, we only need to synthesise Y_δ. This can be done, by using operations from the gate set only:

$$Y_\delta \equiv X_{\pi/2} Z_{\delta+\pi} X_{3\pi/2}$$

In conclusion, since the gate set also provides $\wedge(X)$, any $\wedge(U)$ can be realized through a composition of operations coming from the IBM gate set. □

Not all the existing processors are universal. D-Wave ones are an example. In fact, the company goal is to build specific-purpose processors, meant to explore the field of optimization problems through *quantum annealing* procedures [12].

2.1.3 *Relation Between Classical and Quantum Logic*

It has been shown [13] that, to achieve quantum universality, one can start from an operator universal in the Boolean functions, i.e. the Toffoli, and by adding only a single 1-qubit operator, the operator set is quantum universal. The subject operator can be expressed as $X_{\pi/2} Z_{\pi/2} X_{\pi/2}$.[2] As stated in [14], this "[...] can be interpreted as saying that Fourier transform is really all there is to quantum computation on top of classical", since $X_{\pi/2} Z_{\pi/2} X_{\pi/2}$ corresponds to a Fourier transform.

2.1.3.1 Special Case (i)

Whenever $V_{\alpha,\delta} = Z_\alpha Y_\delta Z_\alpha$, the circuit decomposition simplifies to Fig. 2.9. An important example is $\wedge(Z_\alpha)$ itself: $Z_\alpha = Z_{\alpha/2} \cdot Z_{\alpha/2}$.

2.1.3.2 Special Case (ii)

Whenever $V_{\alpha,\delta} = Z_\alpha Y_\delta Z_\alpha X$, the circuit decomposition simplifies to Fig. 2.10. Important examples are the Pauli matrices Y, Z and X itself.

---

[2] As well as $Y_{\pi/2}X$ or as the more common *Hadamard* gate H. We are not going to use the latter in this thesis as we opted to keep the treating closer to real gate implementations.

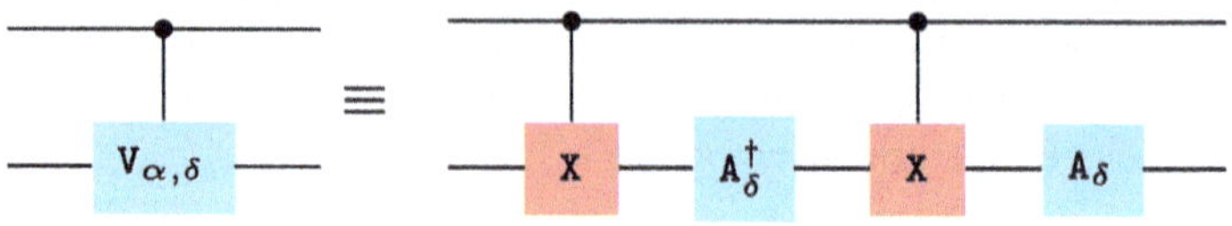

Fig. 2.9 Special case $V_{\alpha,\delta} = Z_\alpha Y_\delta Z_\alpha$

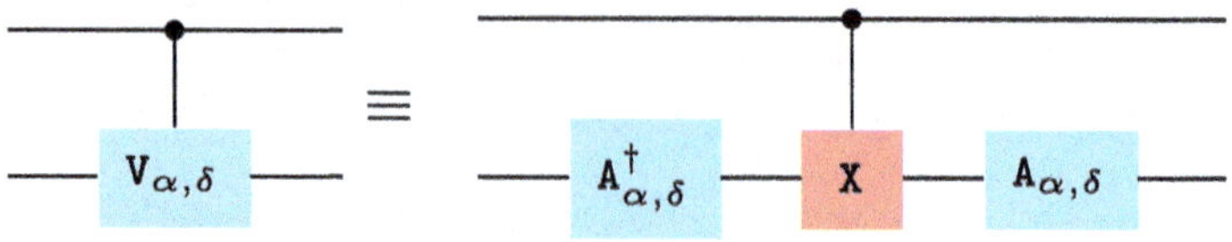

Fig. 2.10 Special case $V_{\alpha,\delta} = Z_\alpha Y_\delta Z_\alpha X$

2.2 Linear Reversible Boolean Functions

A function $g : \mathbb{F}_2^n \rightarrow \mathbb{F}_2^n$ is linear and reversible iff it can be synthesised into a quantum circuit composed by $\wedge(X)$ only. Such a function can be encoded into a squared boolean matrix. A synthesis method finds a matrix decomposition made out of elements M_{uv}; each of which relates to a $\wedge(X)$ gate with control and target qubits indexed by u and v, respectively. Formally, an n-by-n boolean matrix A defines g. A circuit generated by $\wedge(X)$ gates implements a unitary U_g such that:

$$U_g|b\rangle \equiv |g(b)\rangle \equiv |Ab\rangle,$$

for any boolean vector $b \in \mathbb{F}_2^n$.

There exist several methods, each generating a different decomposition [15]. An optimal decomposition is available [16], which is based on *Gaussian elimination*.

This framework can help the designer to build efficient circuits and avoid to produce duplicates.

2.3 The Clifford Group

The Clifford group $\mathbb{C}$ dues its importance to its implication in fault-tolerant computation [17], simulation [18] and benchmarking [19]. Such a group is generated by 3 operators:

$$\mathbb{C} \equiv \langle \wedge(X), X_{\pi/2}, Z_{\pi/2} \rangle. \tag{2.7}$$

$\mathbb{C}$ can be efficiently simulated by a classical computer [20]. This has as comeback that one can evaluate fault-tolerant protocols classically. As drawback, it is not universal. However to achieve universality while at the same time providing an operator set which can realize any quantum evolution efficiently one need to add a single 1-qubit operator, usually assumed to be $R_{\pi/4}$[3] or the corresponding in rotational terms $Z_{\pi/4}$. In fact, any unitary of dimensionality 2 can be approximated efficiently and with arbitrary precision [21–23]. Formally, by properly composing single qubit operators coming from $\mathbb{C}$—i.e. $X_{\pi/2}$, $Z_{\pi/2}$—together with $Z_{\pi/4}$ one can achieve universality in the *dense* sense, by only introducing a polynomial overhead.[4]

Since we are facing with a discrete operator set composed by even fractions of π only, we can re-state the nomenclature. Namely, the same universal operator set can be expressed in the following intuitive way:

$$\mathbb{C}^+ \equiv \langle \wedge(\mathtt{X}), \mathtt{X}^{1/2}, \mathtt{Z}^{1/2}, \mathtt{Z}^{1/4} \rangle. \tag{2.8}$$

This nomenclature stresses the logic behind the relation Pauli-rotational gates, as the power function degree says how many times one needs to apply the rotational gate to simulate a Pauli operator.

2.4 Diagonal Phase Polynomials

Given a boolean $\mathtt{b} \in \mathbb{F}_2^n$, a unitary $\mathtt{D}$ is a diagonal phase polynomial if its action on $|\mathtt{b}\rangle$ can be written as:

$$\mathtt{D}|\mathtt{b}\rangle = e^{\frac{i\pi}{4} f(\mathtt{b})}|\mathtt{b}\rangle, \tag{2.9}$$

with $f : \mathbb{F}_2^n \to \mathbb{F}_8$ being a 3-degree polynomial, composed by three monomial summations [25, 26]—i.e. $f = f_1 + f_2 + f_3$. This function defines the class of algorithms known as *instantaneous quantum polynomials* [27], which has generator set:

$$\{\mathtt{Z}^{1/4}, \wedge(\mathtt{Z}^{1/2}), \wedge_2(\mathtt{Z})\}. \tag{2.10}$$

Notice that applying the $\wedge(\mathtt{Z}^{1/2})$ operator twice results into the *Clifford* gate $\wedge(\mathtt{Z})$. While, applying it three times corresponds to $\wedge(\mathtt{Z}^{-1/2})$.

2.5 Global Gates

We here extend the basis gate set, introducing **global gates**. A global gate is an operation able to entangle several qubits simultaneously [28–32]. For this reason it is a promising resource, both in hardware and software terms. Citing [29]: "It has been suggested that polynomial or exponential speedups can be obtained with global [gates]".

---

[3] Commonly referred as the $\mathtt{T}$ gate.

[4] Other extensions of $\mathbb{C}$ may be of interest. E.g. in Ref. [24] authors consider $\wedge(Z_{\pi/2})$ in their generator set.

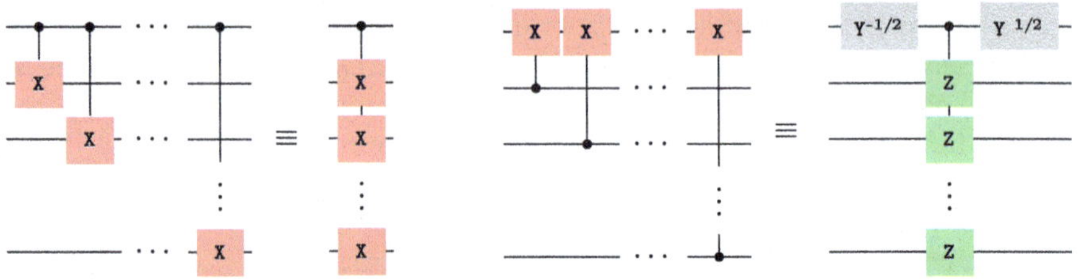

(a) Single control, multiple targets. (b) Single target, multiple controls.

Fig. 2.11 Generic global operations

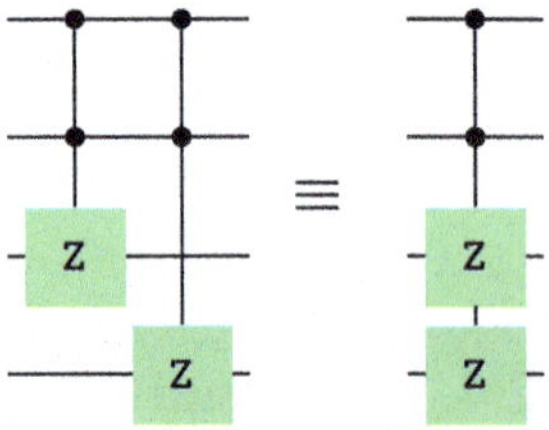

Fig. 2.12 Global gate $\wedge_2(\mathtt{Z} \otimes \mathtt{Z})$

Theoretical advantages can be found in Ref. [33], where authors proved that any n-qubit Clifford circuit can be synthesised to $4n - 6$ global gates and any n-qubit circuit with $\dot{n}$ non-Clifford gates can be synthesised with no more than $2\dot{n} + \mathcal{O}(n/\log n)$ global gates.

In this thesis, we refer to a global gates as an operation featured by either

- *single control, multiple targets* or
- *single target, multiple controls*.

The corresponding circuits are shown in Fig. 2.11. Namely, one can see a global gate as operators $\wedge(\mathtt{X}^{\otimes m})$ or $\wedge(\mathtt{Z}^{\otimes m})$.

Global gates do not limit to the ones presented above. Other kinds may feature multiple control qubits (or target). For our interest there is the $\wedge_2(\mathtt{Z}^{\otimes m})$ operator, which operates globally and features two control qubits. Figure 2.12 shows an example with two control and two target qubits.

As we show within the next Chapter, a *teleportation protocol* implementing a global gate is fast, as the number of qubits involved does not affect the running time. For this reason, computing with global gates on distributed architectures is particularly appealing.

2.6 Transpiling Gates

Usually, a quantum processor does not supplies natively logical gates as $\wedge(\mathtt{X})$ and $\wedge(\mathtt{Z})$. Therefore, there is need for a final mapping process made by the transpiler. A transpiler work with a knowledge base, which can be used to express logical gates in

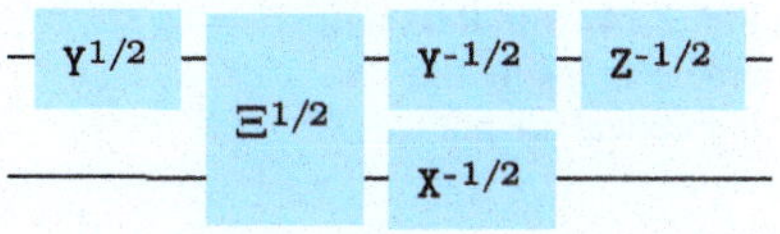

Fig. 2.13 Circuit equivalent to a $\wedge(\mathtt{X})$ gate, using hardware-level operators

terms of native gates. For example, Ion traps work with *Ising gates* and single-qubit rotational gates.

Assuming to work with an ion-trap supplying a two-qubit Ising gate $\Xi^{1/2^m}$ defined as follows:

$$\Xi^{1/2^m} \equiv \cos(\pi/2^{m+1})\mathbb{1} \otimes \mathbb{1} - \mathtt{i}\sin(\pi/2^{m+1})\mathtt{X} \otimes \mathtt{X}$$

the knowledge base will map any $\wedge(\mathtt{X})$ gate as in Fig. 2.13.

Appendix

2.7 Programming in Higher Order Framework

2.7.1 *Time-Ordered Framework*

So far we have implicitly assumed that a quantum evolution undergoes a time-ordered definition. Formally, consider two unitaries $\mathtt{U}$ and $\mathtt{V}$ and a state $|\vartheta\rangle$. Then, any time-ordered framework force us to chose whether $\mathtt{U}$ operates on $|\vartheta\rangle$ before or after $\mathtt{V}$. In circuit representation, these two cases are respectively

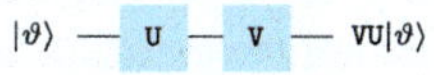

and

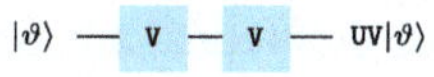

However quantum mechanics allows to think of more general frameworks, where, not only states, but also operators can be superposed to create more complex systems. Such systems may bring new non-classical advantages. Attempts in formally designing a higher-order framework is an active branch of research [34, 35], but this is out of scope. Rather, in the following, we work with a higher-order *oracle* implementing an *indefinite casual order*.

2.7.2 *Indefinite Causal Orders*

The *indefinite causal order* is an interesting property of quantum mechanics. In brief, it is a quantum evolution where two or more operations occur, but the order

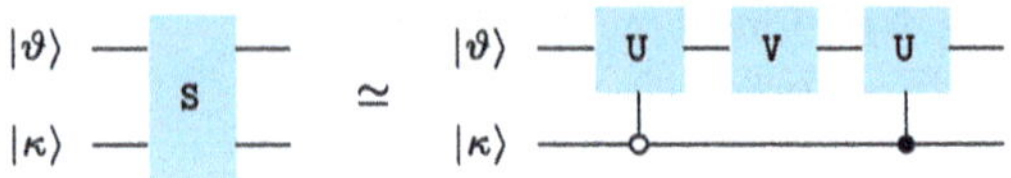

Fig. 2.14 Simulation of oracle S; losing the theoretical advantage

in which they occur is causally ordered by an extra quantum system. This creates a superposition of causal orders among those operations. Such a resource can be used for several purposes. In Refs. [36–38], the indefinite causal orders is considered for thermalization protocols. A big line of investigation deals with enhancing quantum communication [39–42]; we gave the first experimental witness for such a resource [46].[5] As regards computation, indefinite causal order would speed up tasks that involves permuting operations [43–45]. In Ref. [47] authors implement a photonic-based discrimination protocol solved by *superposing unitaries.*

The indefinite causal orders for computation consists essentially on superposing two or more unitaries. The superposition is then coherently conditioned to an auxiliary qubit, called control qubit. For the case of 2 unitaries U, V, the superposing unitary operator S can be defined as follows:

$$\mathtt{S} = \begin{bmatrix} \mathtt{UV} & \mathbf{0} \\ \mathbf{0} & \mathtt{VU} \end{bmatrix}. \tag{2.11}$$

By treating the operator S as an oracle—i.e. it takes a unit of time to run—it brings the advantage of evaluating two different orders at the same time. This advantage comes from the fact that we are extending the standard framework—which originally could only superpose quantum states—to being able to superpose unitaries. This can be used, for example, in Information Processing to distinguish between different evolution by means of a single check [47]. This is unthinkable with standard frameworks where the order of execution of the unitaries must be defined.

However, implementing S does not necessarily reflect the advantage coming from theory. It is necessary to make a distinction to what is a real implementation of S and what is more like a *simulation*. In other words, implementing S means that the physical settings preserve the theoretical advantages. Only in this case one can treat S as an oracle. It is an open question if such a real implementation will ever be possible [48, 49]. In the attempt of finding a solution, a framework meant to superpose *gravitational fields* has been proposed [50].

Stemming from the above, what we can do now is to realize S by means of simulation. Namely, an equivalent evolution which does not preserve the speed-up advantage. An example is shown in Fig. 2.14.

This example immediately shows the theoretical loss, as it requires 2 use for one of the unitaries [51]. Specifically, since U operators run under complementary conditions, one of them does not run, meaning that one time step is always dedicated to perform an identity operation $\mathbb{1}$—an idle time.

[5] We report this in Chap. 4.

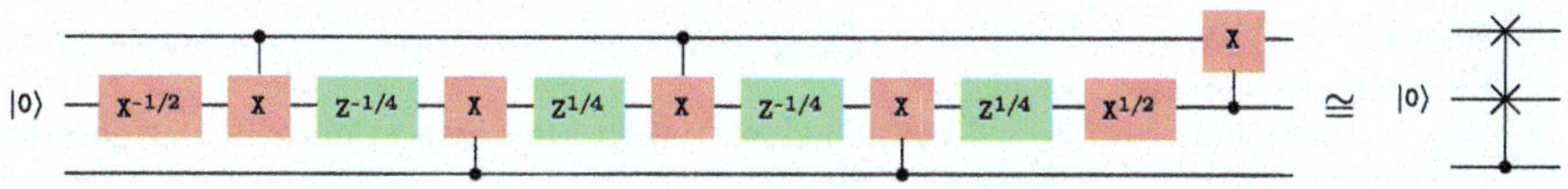

Fig. 2.15 Optimized decomposition of the controlled swap operation

An approach which seems to provide a more natural realization of S is investigated in [45, 52]. These works use auxiliary qubits and controlled swaps. However, to date, there is no technology providing such a gate natively, hence it is necessary to consider it as an oracle to not lose the advantage. The best we can do now is providing an efficient decomposition for it, whenever we have partial knowledge of the input—e.g. an auxiliary qubit being in state $|0\rangle$. Figure 2.15 shows an optimized decomposition w.r.t. the standard one [53].

Unfortunately, as long as native controlled swap are not physically implemented, the speed-up advantage—w.r.t. a time-ordered framework—is still just theoretical and it cannot be witnessed on real implementation. This is especially true when trying to superpose more than two unitaries.

Nevertheless, a non-native implementation of indefinite causal orders may still bring some sort of quantum advantage, as it can be used for *magnitude amplification* [54–56].[6] In Sect. 4.10, we treat this concept—also experimentally—to enhance communication capacity.

References

1. Wernick W (1942) Complete sets of logical functions. J Symb Logic 7(2):99
2. Toffoli T (1980) Reversible computing. In: International colloquium on automata, languages, and programming. Springer, pp 632–644
3. Barenco A (1995) A universal two-bit gate for quantum computation. Proc R Soc Lond. Ser A: Math Phys Sci 449(1937):679–683
4. DiVincenzo DP (1995) Two-bit gates are universal for quantum computation. Phys Rev A 51(2):1015
5. Feynman RP (1985) Quantum mechanical computers. Opt News 11(2):11–20
6. Barenco A et al (1995) Elementary gates for quantum computation. Phys Rev A 52(5):3457
7. IBM. IBM Quantum backends. https://github.com/Qiskit/ibmq-device-information/tree/master/backends. [Online; accessed 14 Nov 2022]
8. IBM. A high-fidelity, two-qubit cross-resonance gate using interference couplers. https://research.ibm.com/publications/a-high-fidelity-two-qubit-cross-resonance-gate-using-interference-couplers. [Online; accessed 29 Nov 2022]
9. Qiskit. The cross-resonance gate for superconducting qubits implements a $Z \otimes X$ interaction. https://qiskit.org/documentation/stubs/qiskit.circuit.library.RZXGate.html. [Online; accessed 29 Nov 2022]
10. Rigetti C, Devoret M (2010) Fully microwave-tunable universal gates in superconducting qubits with linear couplings and fixed transition frequencies. Phys Rev B 81(13):134507
11. McKay DC et al (2017) Efficient Z gates for quantum computing. Phys Rev A 96(2):022330

[6] Useful, e.g., to implement Grover's algorithm oracle [57].

12. Waver D (2022) What is quantum annealing? https://docs.dwavesys.com/docs/latest/c_gs_2.html. [Online; accessed 14 Nov 2022]
13. Shi Y (2003) Both Toffoli and controlled-NOT need little help to do universal quantum computing. Quantum Inf Comput 3(1):84–92
14. Aharonov D (2003) A simple proof that Toffoli and Hadamard are quantum universal. arXiv preprint quant-ph/0301040
15. De Brugiere Timothee G et al (2021) Reducing the depth of linear reversible quantum circuits. IEEE Trans Quantum Engin 2:1–22
16. Patel KN, Markov IL, Hayes JP (2008) Optimal synthesis of linear reversible circuits. Quantum Inf Comput 8(3):282–294
17. Anderson JT, Duclos-Cianci G, Poulin D (2014) Fault-tolerant conversion between the steane and reed-muller quantum codes. Phys Rev Lett 113(8):080501
18. Gottesman D (1998) Theory of fault-tolerant quantum computation. Phys Rev A 57(1):127
19. Magesan E, Gambetta JM, Emerson J (2012) Characterizing quantum gates via randomized benchmarking. Phys Rev A 85(4):042311
20. Gidney C (2021) Stim: a fast stabilizer circuit simulator. Quantum 5:497
21. Dawson CM, Nielsen MA (2005) The solovay-kitaev algorithm. arXiv preprint quant-ph/0505030
22. Kliuchnikov V (2013) Synthesis of unitaries with Clifford+ T circuits. arXiv preprint arXiv:1306.3200
23. Vatan F, Williams C (2004) Optimal quantum circuits for general two-qubit gates. Phys Rev A 69(3):032315
24. Glaudell AN, Ross NJ, Taylor JM (2021) Optimal two-qubit circuits for universal fault-tolerant quantum computation. npj Quantum Inf 7(1):1–11
25. Campbell ET, Howard M (2017) Unifying gate synthesis and magic state distillation. Phys Rev Lett 118(6):060501
26. Cowtan A et al (2019) Phase gadget synthesis for shallow circuits. arXiv preprint arXiv:1906.01734
27. Bremner MJ, Montanaro A, Shepherd DJ (2017) Achieving quantum supremacy with sparse and noisy commuting quantum computations. Quantum 1:8
28. Maslov D, Nam Y (2018) Use of global interactions in efficient quantum circuit constructions. New J Phys 20(3):033018
29. Lu Y et al (2019) Global entangling gates on arbitrary ion qubits. Nature 572(7769):363–367
30. Casanova J et al (2012) Quantum simulation of interacting fermion lattice models in trapped ions. Phys Rev Lett 108(19):190502
31. Ivanov SS, Ivanov PA, Vitanov NV (2015) Efficient construction of three-and four-qubit quantum gates by global entangling gates. Phys Rev A 91(3):032311
32. Martinez EA et al (2016) Compiling quantum algorithms for architectures with multi-qubit gates. New J Phys 18(6):063029
33. van de Wetering J (2021) Constructing quantum circuits with global gates. New J Phys 23(4):043015
34. Chiribella G, Mauro D'Ariano G, Perinotti P (2008) Quantum circuit architecture. Phys Rev Lett 101(6):060401
35. Chiribella G, Kristjánsson H (2019) Quantum Shannon theory with superpositions of trajectories. Proc R Soc A 475(2225):20180903
36. Felce D, Vedral V (2020) Quantum refrigeration with indefinite causal order. Phys Rev Lett 125(7):070603
37. Simonov K et al (2022) Work extraction from coherently activated maps via quantum switch. Phys Rev A 105(3):032217
38. Guha T, Alimuddin M, Parashar P (2020) Thermodynamic advancement in the causally inseparable occurrence of thermal maps. Phys Rev A 102(3):032215
39. Ebler D, Salek S, Chiribella G (2018) Enhanced communication with the assistance of indefinite causal order. Phys Rev Lett 120(12):120502

40. Salek S, Ebler D, Chiribella G (2018) Quantum communication in a superposition of causal orders. arXiv preprint arXiv:1809.06655
41. Caleffi M, Cacciapuoti AS (2020) Quantum switch for the quantum internet: noiseless communications through noisy channels. IEEE J Select Areas Commun 38(3):575–588
42. Koudia S et al (2022) How deep the theory of quantum communications goes: superadditivity, superactivation and causal activation. In: IEEE communications surveys and tutorials
43. Chiribella G (2012) Perfect discrimination of no-signalling channels via quantum superposition of causal structures. Phys Rev A 86(4):040301
44. Araújo M, Costa F, Brukner Č (2014) Computational advantage from quantum-controlled ordering of gates. Phys Rev Lett 113(25):250402
45. Colnaghi T et al (2012) (2010) Quantum computation with programmable connections between gates. Phys Lett A 376(45):2940–2943
46. Cuomo D, Caleffi M, Cacciapuoti AS (2021) Experiencing the communication advantage of the superposition of causal orders. In: IEEE 22nd international workshop on signal processing advances in wireless communications (SPAWC). IEEE, pp 181–185
47. Procopio LM et al (2015) Experimental superposition of orders of quantum gates. Nat Commun 6(1):1–6
48. Vilasini V, Renner R (2022) Embedding cyclic causal structures in acyclic spacetimes: no-go results for process matrices. arXiv preprint arXiv:2203.11245
49. Rubino G et al (2017) Experimental verification of an indefinite causal order. Sci Adv 3(3):e1602589
50. Paunković N, Vojinović M (2020) Causal orders, quantum circuits and spacetime: distinguishing between definite and superposed causal orders. Quantum 4:275
51. Chiribella G et al (2013) Quantum computations without definite causal structure. Phys Rev A 88(2):022318
52. Renner MJ, Brukner Č (2022) Computational advantage from a quantum superposition of qubit gate orders. Phys Rev Lett 128(23):230503
53. Shende VV, Markov IL (2009) On the CNOT-cost of TOFFOLI gates. Quantum Inf Comput 9(5):461–486
54. Zavatta A, Fiurášek J, Bellini M (2011) A high-fidelity noiseless amplifier for quantum light states. Nat Photon 5(1):52–56
55. Kim MS et al (2008) Scheme for proving the bosonic commutation relation using single-photon interference. Phys Rev Lett 101(26):260401
56. Zavatta A et al (2009) Experimental demonstration of the bosonic commutation relation via superpositions of quantum operations on thermal light fields. Phys Rev Lett 103(14):140406
57. Zalka C (1999) Grover's quantum searching algorithm is optimal. Phys Rev A 60(4):2746

Chapter 3
Entanglement-Based Computation

Entanglement stands as one of the most captivating phenomena arising from the principles of quantum mechanics. It also emerges as a highly promising resource in the field of quantum science.

To our purpose we focus on one of the four Bell states. These states are fully entangled, meaning that their correlation is maximal and the system is said to be close. Let us introduce the $|\Phi^+\rangle$ state, defined as follow:

$$|\Phi^+\rangle = \frac{1}{\sqrt{2}}(|00\rangle + |11\rangle).$$

Since the two system in $|\Phi^+\rangle$ present a *non-local* correlation, this state can be used to perform non-local operations. Physically speaking, this means that one need to perform what is called *entanglement generation and distribution* [1–3]:

- generation; despite the non-local correlation, the generation happens between system which are in proximity one another. Generating a maximally entangled state with high fidelity is generally hard and time-consuming.
- distribution; once that the entanglement is ready, it is possible to relocate the two systems. The entanglement is, in principle, preserved.

As mentioned in the previous chapter, as we proceed into the design a distributed quantum computing *ecosystem*, some assumptions are necessary to give insights and performance analysis characterizing our design. We assume a system should provide a logical primitive, $\mathfrak{E}$, which can be used as a reliable resource for computation. In the circuit formalism, this correspond to an operator with no inputs and two output wires:

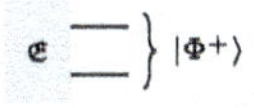

D. Cuomo, *Architectures and Circuits for Distributed Quantum Computing*, Springer Theses, https://doi.org/10.1007/978-3-031-73808-1_3

Below we outline some of the most promising resources for distributed quantum computing, all exploiting $\mathfrak{E}$. Before proceeding, we need to introduce a formalism for *measurement-based computation*.

3.1 Measurement-Based Computation

A computing paradigm is called measurement-based whenever the quantum computation is interleaved by measurements acting on a sub-system [4]. The output of a measurement is then used to perform classical conditioned quantum operations.

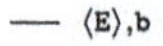

The output can then be used to choose whether performing or not a unitary operation U over another qubit:

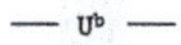

Whenever to a Pauli gate follows a Pauli measurement, there is no need to apply the former. Precisely, instead of applying the quantum gate followed by the measurement, one can always perform the measurement and then applying a classical correction on the output, corresponding to the Pauli gate. Such a technique may be referred as *pushing* technique as, intuitively, one can picture this manipulation as pushing the gate beyond the measurement. See circuits in Fig. 3.1 for an example.

3.2 Teleportation

The first resource we report is called *teleportation* [3]. It is a protocol that, by fact, teleport quantum information from a system to another, by means of entanglement. Figure 3.2 shows the protocol steps in circuit representation. The wires are grouped by color, representing different processor, or memories. A generic processor is referred as p_i. In the case of teleportation 2 processors p_i, p_j are involved, which has the roles of *sender* and *receiver*. At the beginning of the protocol, the quantum information $|\varphi\rangle$ is located within the sender, together with half of the Bell state $|\Phi^+\rangle$. The receiver need to store the other part of the Bell state. At this point the receiver need to wait that the receiver perform a few operations meant to "inject" the information within

Fig. 3.1 Pushing technique to avoid quantum gates

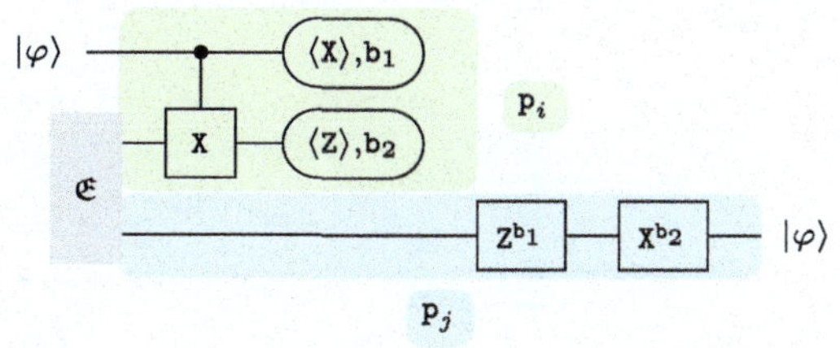

Fig. 3.2 Quantum teleportation protocol between two parties

its half part of the entangled state. Because of the entanglement, now the quantum information is already spread all over the system, which means that $\mathtt{p}_j$, the receiver, also has partial knowledge of it.

The end of this part consists on measuring the entire system in the orthogonal basis $\mathtt{X} \otimes \mathtt{Z}$. This is a necessary step in order to ensure that the receiver will get exactly $|\varphi\rangle$. In fact, the measurement will produce two boolean values, $\mathtt{b}_1, \mathtt{b}_2$, that the receiver will need to correct its state.

To understand how such a resource can be used in computation, consider that it is often the case where two states need to interact but they are stored in such a position that does not allow them to do so, unless some middle step is performed to approach them. This means that a *routing* protocol would take care of moving those states up to a couple of qubits which are able to interact one another. This can be done by means of teleportation [5]. However, in local quantum computation, it is more common to find some smart criteria to re-arrange the state storage when necessary, by means of *swapping* protocols [6–8]. Even if this is the most common approach, it may be smart as well to instead consider teleportation as an alternative to swapping protocols.

When considering distributed architecture, this are generally assumed to deeply exploit entanglement for their interconnection. Hence, it is more likely to see in the future proposal exploiting teleportation, rather than swapping.

3.3 Non-Local Operations

The teleportation protocol results to be a basic example of a wider class of teleporting protocols. It is in fact possible to, not only teleport states, but entire operations. For this reason the procedure we report here can be also referred as *tele-gate*.

We already discussed the importance of the $\wedge(\mathtt{X})$ operator for quantum computation in Sect. 2.1. We here report a way to perform such an operation between states belonging to different processors by means of non-local operations. We here report a standard protocol to perform a non-local $\wedge(\mathtt{X})$, employing a few steps; as shown in Fig. 3.3.

Notice that performing a non-local operation is not limited to the consumption of a Bell state $|\Phi^+\rangle$. Rather, as discussed in Sect. 5.5, one can use any Bell state. For the sake of simplicity, we will always refer to $|\Phi^+\rangle$ states.

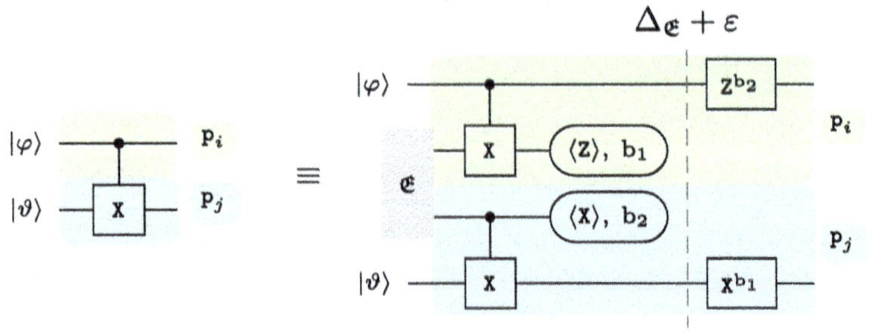

Fig. 3.3 Tele-gate performing $\wedge(\mathtt{X})$ between qubits belonging different processors

Similarly to the teleportation protocol—see Fig. 3.2—, there is a step meant two inject the control and target states within the entangled state. Thanks to this step, some quantum information is exchanged between the two parties. However, to ensure the equivalence with the subject operator, a measurement step is necessary. Hence, the state that was originally fully entangled in $|\Phi^+\rangle$ is measured in the orthogonal basis $\mathtt{Z} \otimes \mathtt{X}$. The output $\mathtt{b}_1, \mathtt{b}_2$ is then subject to a cross communication through classical channels and eventually used to perform Pauli corrections, i.e., the last step of the procedure: $\mathtt{Z}^{\mathtt{b}_2}$ over the control qubit and $\mathtt{X}^{\mathtt{b}_1}$ over the target qubit.

Mindful of Sect. 2.1, we don't need further non-local operations in order to achieve universal computation. However, it is useful to know that other tele-gates are certainly possible. E.g. Fig. 3.4 shows the procedure to perform a $\wedge(\mathtt{Z})$ non-local operator.

One last useful observation concerns how to relate the protocol run-time with the discrete time domain space we defined in Sect. 1.3.1. In the previous section, we refer to a unit of time as being, *roughly*, the time to generate a link. Let this quantity be $\Delta_{\mathfrak{E}}$. Forgetting, for a moment, about the post-processing, there is a short delay ε caused by local operations. We argue that such a quantity is realistic since (i) two local operations are generally much faster than $\Delta_{\mathfrak{E}}$ and (ii) as we proceed to define more complex protocols, ε remains invariant.

Fig. 3.4 Tele-gate performing $\wedge(\mathtt{Z})$ between qubits belonging to different processors

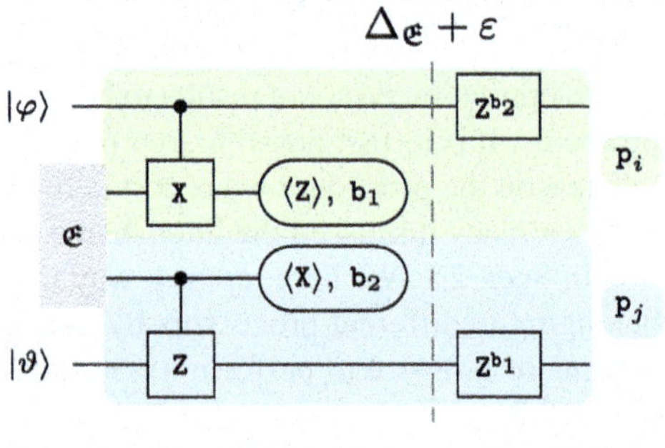

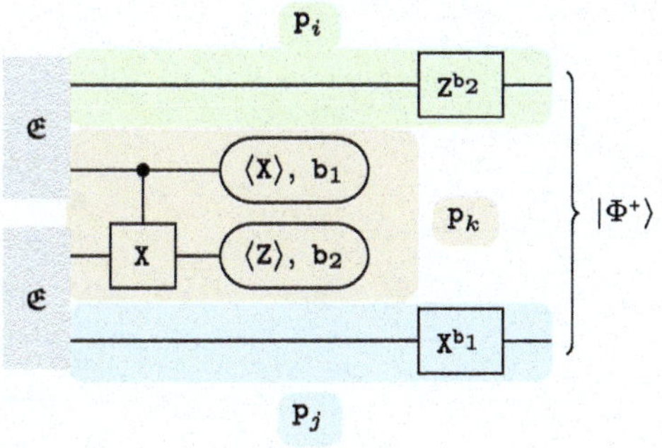

Fig. 3.5 The entanglement swap protocol inter-connects two processors by means of an intermediate one, which has direct connection with both of them

3.4 Entanglement Swap

Here we report a protocol known as *entanglement swap*. It is a promising procedure as it models scalable distributed architecture. Figure 3.5 shows the parties involved and their actions. Specifically, assume that two processors p_i, p_j cannot rely on a direct inter-connection through an entanglement pair. However they both share an interconnection with a *middle* processor p_k. The middle processor has therefore stored in his memory two half of entangled pairs. By means of a local $\wedge(X)$ on his system, p_k inter-connects p_i with p_j. As usual, an orthogonal measurement $X \otimes Z$ is necessary to apply eventual corrections over p_i and p_j, which at the end of the procedure share a fully entangled state, ensuring the new inter-connection.

Depending on the time model adopted when dealing with this procedure, the inter-connectivity among the processors find different treating. For example, the work done in [9] has a more dynamic-like approach, making a distinction between *link* and *virtual link*. Such a choice probably comes from an interest in modeling a network of quantum technologies, where it is more common to deal with *online* combinatorial problems [10]. Instead, our focus here is to smartly model distributed architecture meant to perform algorithms. This brings us to see the inter-connectivity within a more static time model; which translates into a simpler modeling for the connectivity. We explain this in detail within next Sect. 3.5.

3.5 Entanglement Path

As our focus is on the treating of distributed quantum computing, we here provides a non-local $\wedge(X)$, which makes use of entanglement swaps. The circuit in Fig. 3.6 comes from the combination of the basic implementation of a non-local $\wedge(X)$—see Sect. 3.3—with the entanglement swap protocol.

It is important to notice that all the measurements happen at the same time. Hence, the ε delay is constant as anticipated. Even more important to know is that this result can be generalized to any number of middle processors $p_{k_1}, p_{k_2}, \ldots, p_{k_m}$. For

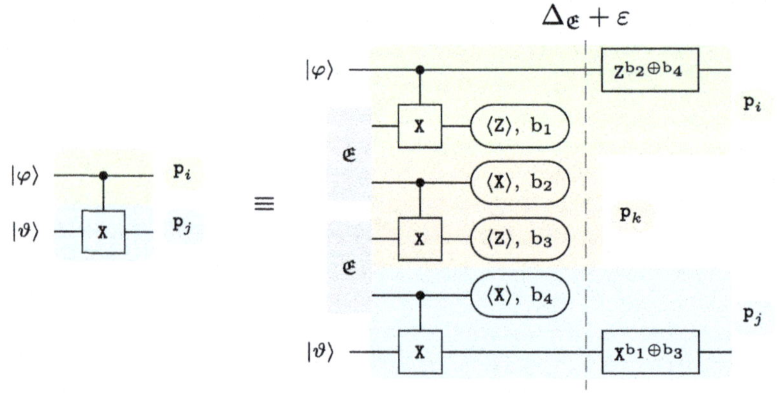

Fig. 3.6 Tele-gate $\wedge(X)$ by means of entanglement swap

this reason we refer to $\{p_i, p_{k_1}, p_{k_2}, \ldots, p_{k_m}, p_j\}$ as an *entanglement path* of length $m + 1$. We now give an inductive proof for this result.

Theorem 3.1 *An entanglement path* $\{p_{i_1}, p_{i_2}, \ldots, p_{i_m}\}$ *has an implementation with depth 3.*

Proof Consider an entanglement path of length 2. A naive realization consists on putting in strict sequence two entanglement swaps:

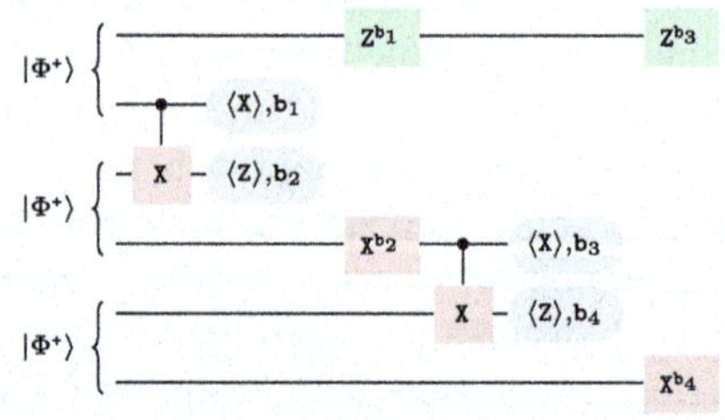

Pauli gates are the only ones we are going to optimize; since the others are independent and no optimization can be applied. What follows is the base case for the induction:

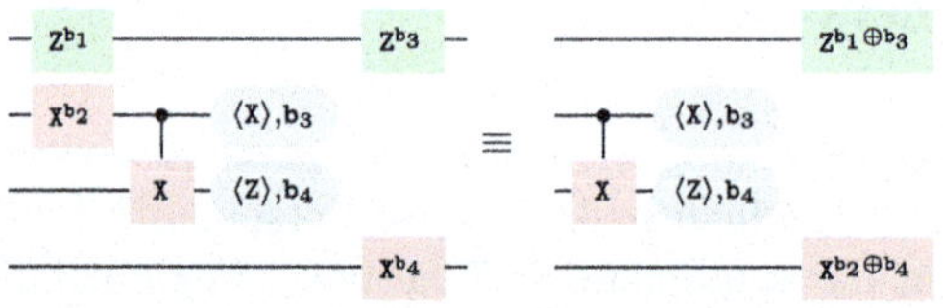

The r.h.s. of the above equation has post-processing composed by $Z^{b_1 \oplus b_3}$ on first qubit and $X^{b_2 \oplus b_4}$ on last qubit. Notice that the measurements are independent from other operations.

By assuming that such a shape is preserved in the inductive step, we show that this transformation can be applied to any length m:

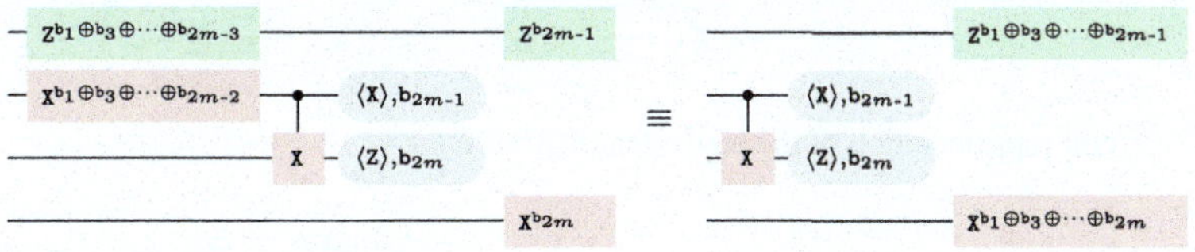

This proves that we can always consider an entanglement path $\{p_{i_1}, p_{i_2}, \ldots, p_{i_m}\}$ to have circuit depth 3.

We just showed an efficient implementation for the entanglement path. Now we do one last step to exploit such a result and performing a generalized remote operation efficiently.

Theorem 3.2 *A tele-gate of entanglement path* $\{p_{i_1}, p_{i_2}, \ldots, p_{i_{m+2}}\}$ *has depth 3.*

Proof The theorem above allows us to assume that, to perform a remote operation by using a path of length m, the computing qubits interact only with two communications qubits and depend only by Pauli operations $Z^{b_1 \oplus b_3 \oplus \cdots \oplus b_{2m-1}}$ and $X^{b_2 \oplus b_4 \oplus \cdots \oplus b_{2m}}$. We can further *propagate* such operations as follows:

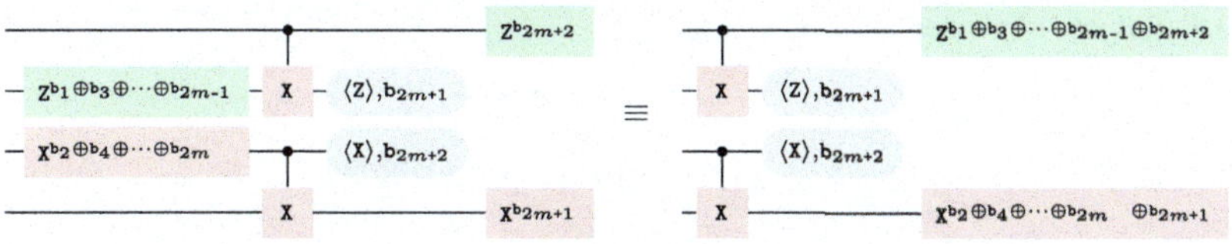

In this way the measurements are independent and the depth of the circuit is not increased.

3.6 Entanglement Tree

Driven again by the aim of finding smart strategies to perform non-local gates with low consumption of entanglement links; we now investigate a way to perform multiple operations **by means of the same link**. To get a first intuition of what we are going to show, consider the following equivalence:

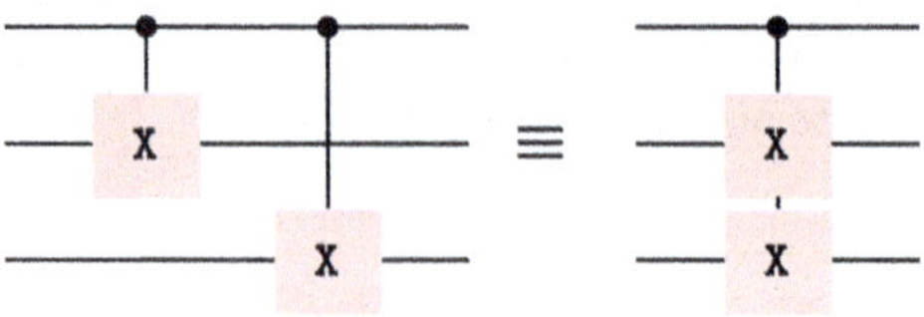

Fig. 3.7 Circuit representation of the equivalence (3.1)

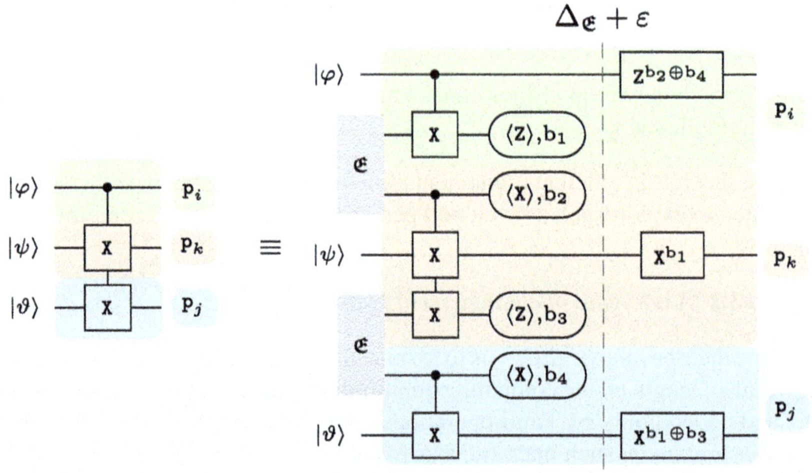

Fig. 3.8 Remote implementation of $\wedge(\mathrm{X} \otimes \mathrm{X})$

$$\wedge\,(\mathbb{1} \otimes \mathrm{X}) \cdot \wedge(\mathrm{X} \otimes \mathbb{1}) \equiv \wedge(\mathrm{X} \otimes \mathrm{X}); \tag{3.1}$$

which has circuit representation reported in Fig. 3.7.

Beyond its simplicity, equivalence (3.1) gives us a different perspective to implement multiple non-local operations by means of few entanglement links. This happens for the case of two processors p_i, p_j, but it can be generalized. In fact, a corollary that follows from Theorem 3.2 and from [11, 12] is that a global gate $\wedge(\mathrm{X}^{\otimes m})$—or $\wedge(\mathrm{Z}^{\otimes m})$—, distributed over a network $\mathcal{Q} = (\mathfrak{p}, \mathfrak{l})$, involves a sub-set of links forming what we call an **entanglement tree**. Namely, a sub-graph satisfying the standard properties of a tree.

In fact, stemming from Eq. (3.1), the target systems are essentially independent, up to the common control qubit. Hence, there is no reason to restrict them to be part of the same processor. For the case under consideration—i.e. $\wedge(\mathrm{X} \otimes \mathrm{X})$—, the maximum number of processors is three. Figure 3.8 shows the circuit protocol to perform $\wedge(\mathrm{X} \otimes \mathrm{X})$, where the system is spread over three processors.

We will use this technique in Chap. 5 to minimize entanglement links consumption.

3.7 The Role of Clifford Group in Distributed Architectures

In our model, we showed that by postponing the Pauli-corrections, we get the combined advantage of (i) parallelizing remote operations and (ii) delaying the correction, which amortizes the impact of the traveling time that a boolean value takes to reach its destination(s). An ideal result would be to push all the corrections to the end of the circuit. In fact, as already discussed in this chapter, if the corrections reach the end of the circuit, they could be replaced by classical computation. Driven by this goal, we now investigate the properties of quantum circuits to find when such a condition is satisfied, starting from the Clifford group $\mathbb{C}$, restated as follows:

$$\mathbb{C} \equiv \langle \wedge(\mathtt{X}), \wedge(\mathtt{Z}), \mathtt{X}^{1/2}, \mathtt{Z}^{1/2}, \mathtt{Y}^{1/2} \rangle$$

Even though the Clifford group is not universal, its properties make it a good starting point for the design of a computational paradigm for distributed architectures. For example, adding only one extra operator to the group generator makes it universal. In Sect. 2.3, we referred to such an extension as the group $\mathbb{C}^+ \equiv \langle \wedge(\mathtt{X}), \mathtt{X}^{1/2}, \mathtt{Z}^{1/2}, \mathtt{Z}^{1/4} \rangle$.

3.7.1 *Implication on Post-processing*

As said at the beginning of Sect. 2.3, important benefits could be achieved by postponing the Pauli corrections to the end of the circuit, where they can be computed classically. In the context of Clifford circuits, the distributed computation results to be independent from classical communication, as we can always apply the following rules:

- $\wedge(\mathtt{X}) \cdot \mathtt{X}^b \otimes \mathbb{1} \equiv \mathtt{X}^b \otimes \mathtt{X}^b \cdot \wedge(\mathtt{X})$
- $\wedge(\mathtt{X}) \cdot \mathbb{1} \otimes \mathtt{Z}^b \equiv \mathtt{Z}^b \otimes \mathtt{Z}^b \cdot \wedge(\mathtt{X})$
- $[\wedge(\mathtt{X}), \mathbb{1} \otimes \mathtt{X}^b] = [\wedge(\mathtt{X}), \mathtt{Z}^b \otimes \mathbb{1}] = 0.$

Similarly, for $\mathcal{L}_{\wedge(\mathtt{Z})}$ circuits, we can use the following rules:

- $\wedge(\mathtt{Z}) \cdot \mathtt{X}^b \otimes \mathbb{1} \equiv \mathtt{X}^b \otimes \mathtt{Z}^b \cdot \wedge(\mathtt{Z})$
- $[\wedge(\mathtt{Z}), \mathtt{Z}^b \otimes \mathbb{1}] = [\wedge(\mathtt{Z}), \mathbb{1} \otimes \mathtt{Z}^b] = 0$

Finally, the last single layer circuit $\mathcal{L}_{\mathtt{Y}^{1/2}}$ can be handled as follows:

- $\mathtt{Y}^{1/2} \cdot \mathtt{X}^b \cong \mathtt{Z}^b \cdot \mathtt{Y}^{1/2}$
- $\mathtt{Y}^{1/2} \cdot \mathtt{Z}^b \equiv \mathtt{X}^b \cdot \mathtt{Y}^{1/2}$

- $\mathtt{X}^{1/2} \cdot \mathtt{Z}^b \cong \mathtt{Z}^b\mathtt{X}^b \cdot \mathtt{X}^{1/2}$
- $\mathtt{Z}^{1/2} \cdot \mathtt{X}^b \cong \mathtt{X}^b\mathtt{Z}^b \cdot \mathtt{Z}^{1/2}$
- $[\mathtt{X}^{1/2}, \mathtt{X}^b] = [\mathtt{Z}^{1/2}, \mathtt{Z}^b] = 0.$

Remark 3.1 By means of the above rules, all the post-processing operations can be pushed forward, up to end of the circuit and they can be computed efficiently by a classical computer. Furthermore, since no post-processing occurs during quantum computation, the entanglement path length has negligible impact to the running-time (thanks to the non-locality of the operations).

3.8 Compiling and Scheduling a Circuit

The results shown in Sect. 3.7.1 unveil the dependency of quantum computation from classical communication. We can characterize computation with a simple model, that rules out the specifics of classical channels. Specifically, for a given network $\mathcal{Q} = (\mathfrak{p}, \mathfrak{l})$, we already defined its dynamics as time-steps $\mathfrak{t}_1, \mathfrak{t}_2, \ldots$. Each of which relates to some subset $\mathcal{Q}_\mathfrak{t} \subseteq \mathcal{Q}$. We want to keep consistency between network and circuit dynamics and, for this reason, we re-state the concept of *layer*[1] to one more suitable for our scenario. Let $\ell_\mathfrak{t}$ be the set of operations running within time-step $\mathfrak{t}$. We refer to $\ell_\mathfrak{t}$ the layer running at time $\mathfrak{t}$, a compiler is a map $\mathcal{C} \mapsto \mathcal{L}$ with $\mathcal{C}$ being a quantum circuit and

$$\mathcal{L} \equiv \{\ell_\mathfrak{t}\}_\mathfrak{t}.$$

A compiled circuit $\mathcal{L}$ will pass by a scheduler and a transpiler, which will take care of allocating resources to run $\mathcal{L}$. Specifically, for each $\ell_\mathfrak{t} \in \mathcal{L}$ the scheduler builds a subset $\mathcal{Q}_\mathfrak{t} \subseteq \mathcal{Q}$. The optimization ends with a transpiler, which maps logical gates to physical ones. The *quality* of distributed computation is given by:

- the $\mathfrak{E}$-depth, which is the cardinality $|\mathcal{L}|$, and
- the $\mathfrak{E}$-count, which is the overall number of links used to perform each layer $\ell_\mathfrak{t} \in \mathcal{L}$.

The transpiler very depends on the specifics of each processor, hence, it is out of the scope of this thesis.

3.9 Protocols for Universal Computation

We now identify groups of circuits that are pivotal in the context of universal computation. We will work with such groups to provide a simple and rigorous compilation model.

[1] A circuit layer commonly refers to a set of gates that appears within the same column.

3.9.1 Controlled Unitary Protocol

The strategy we show here, introduced in [13, 14], is another way of performing non-local operations. It is quite similar and, for many scenarios, completely equivalent to the protocols we introduced this far. However, features coming either from hardware or algorithm would happen to benefit from this slightly different strategy. For our example, consider the gate $\wedge(\mathtt{U})$. As we are facing a parameterized target unitary, Pauli correction do not propagate trivially in general. For this reason, it may be better to apply the correction *before* the gate, as showed in Fig. 3.9. Notice that this approach forces processor $\mathtt{p}_j$ to wait for the value $\mathtt{b}_1$, coming from $\mathtt{p}_i$, before proceeding with the computation.

3.9.2 Implications on Protocol's Run-Time

As mentioned, Pauli operators propagate non-trivially over more general gates—e.g. the controlled-phase gate $\wedge(\mathtt{Z}^{1/2})$—, we are forced to apply the corrections *before* the local gate. This means that local gates are delayed, waiting for the arrival of classical information. Nevertheless, it is important to notice that such a delay does not depend on the number of qubits involved in the operation, resulting again in efficient implementation of global gates. See, as an example, the protocol in Fig. 3.10.

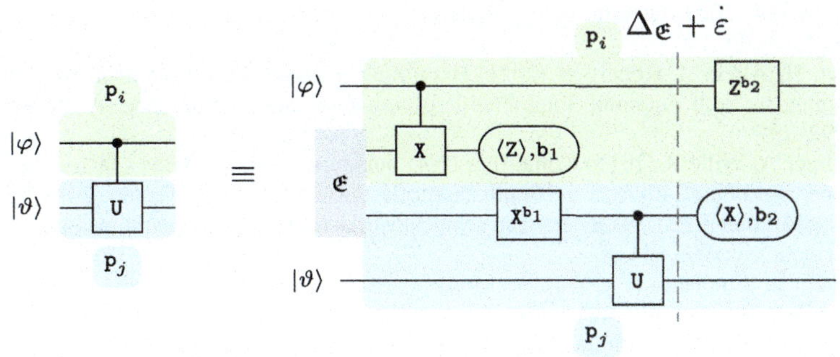

Fig. 3.9 Non-local implementation of $\wedge(\mathtt{U})$

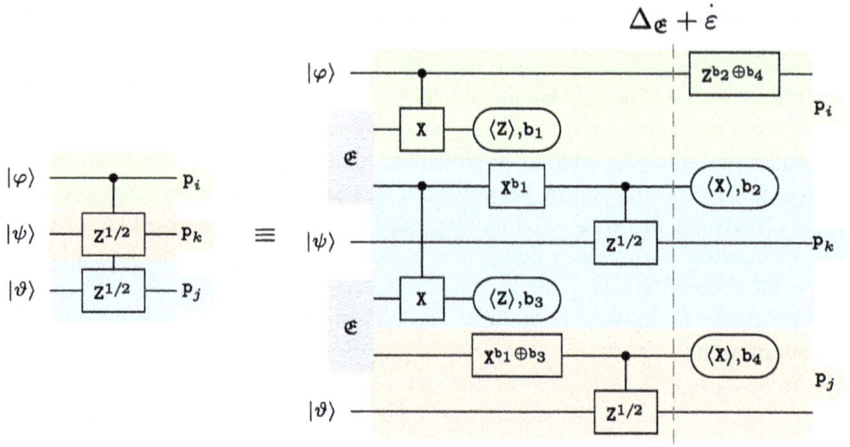

Fig. 3.10 Remote implementation of $\wedge(Z^{1/2} \otimes Z^{1/2})$

References

1. Cacciapuoti AS et al (2019) Quantum internet: networking challenges in distributed quantum computing. IEEE Netw 34(1):137–143
2. Cacciapuoti AS et al (2020) When entanglement meets classical communications: quantum teleportation for the quantum internet. IEEE Trans Commun 68(6):3808–3833
3. Cuomo D, Caleffi M, Cacciapuoti AS (2020) Towards a distributed quantum computing ecosystem. IET Quantum Commun 1(1):3–8
4. Nielsen MA (2023) Quantum computation by measurement and quantum memory. Phys Lett A 308(2–3):96–100
5. Stefan H, Alwin Z, Robert W (2021) Exploiting quantum teleportation in quantum circuit mapping. In: 26th Asia and South Pacific design automation conference (ASP-DAC). IEEE, pp 792–797
6. Zulehner A, Wille R (2019) Compiling $U(4)$ quantum circuits to IBM QX architectures. In: Proceedings of the 24th Asia and South Pacific design automation conference 2019, pp 185–190
7. O'Gorman B et al (2019) Generalized swap networks for near-term quantum computing. arXiv preprint arXiv:1905.05118
8. Madden L, Simonetto A (2022) blackBest approximate quantum compiling problems. ACM Trans Quantum Comput 3(2):1–29
9. Davide F et al (2021) Compiler design for distributed quantum computing. IEEE Trans Quantum Engin 2:1–20
10. Van Hentenryck P, Bent R (2006) Online stochastic combinatorial optimization. The MIT Press
11. Andres-Martinez P, Heunen C (2019) Automated distribution of quantum circuits via hypergraph partitioning. Phys Rev A 100(3):032308
12. Yimsiriwattana A, Lomonaco SJ Jr (2004) Generalized GHZ states and distributed quantum computing. arXiv preprint quant-ph/0402148 (2004)
13. Eisert J et al (2000) Optimal local implementation of nonlocal quantum gates. Phys Rev A 62(5):052317
14. Jiang L et al (2007) Distributed quantum computation based on small quantum registers. Phys Rev A 76(6):062323

Chapter 4
Essentials on Quantum Noise

4.1 Quantum Noise

Assuming a good characterizing model for the noise affecting quantum information can be challenging. A highly generic one can be described as the evolution $\mathcal{N}$ on an d-dimensional system, where a state of interest σ evolves together with the environment in a state ρ [1]:

$$\mathcal{N}(\sigma) = \mathtt{Tr}_{\mathtt{env}}\big(\mathrm{U}(\sigma \otimes \rho)\mathrm{U}^{\dagger}\big).$$

Let $\{|e_k\rangle\}_k$ be an orthonormal basis for the environment. One can assume the environment system as being in the state $\rho = |e_0\rangle\langle e_0|$. This assumption is non-restrictive as the basis is generic and if the considered environment was not pure, one can always introduce an extra *reference* system to purify ρ. It follows that

$$\mathcal{N}(\sigma) = \sum_k \langle e_k| \mathrm{U}(\sigma \otimes |e_0\rangle\langle e_0|)\mathrm{U}^{\dagger} |e_k\rangle .$$

By defining $\mathrm{N}_k = \langle e_k| \mathrm{U} |e_0\rangle$, known as *Kraus* operator, $\mathcal{N}$ becomes

$$\mathcal{N}(\sigma) = \sum_k \mathrm{N}_k \sigma \mathrm{N}_k^{\dagger}. \tag{4.1}$$

Decoherence is the most problematic noise affecting information. This can be represented with the Kraus formalism in terms of Pauli operators acting on each qubit independently [2–4], hence the set $\{\mathrm{N}_k\}_k$ as the form $\{\sqrt{\mathrm{p}_k}\,\mathrm{E}_k\}_k$, such that $\sum_k \mathrm{p}_k = 1$ and E_k belongs to the *Pauli group* $\mathbb{P} \equiv \{\pm\mathbb{1}, \pm\mathtt{i}\mathbb{1}, \pm\mathtt{X}, \pm\mathtt{iX}, \pm\mathtt{Y}, \pm\mathtt{iY}, \pm\mathtt{Z}, \pm\mathtt{iZ}\}^{\otimes d}$.

D. Cuomo, *Architectures and Circuits for Distributed Quantum Computing*,
Springer Theses, https://doi.org/10.1007/978-3-031-73808-1_4

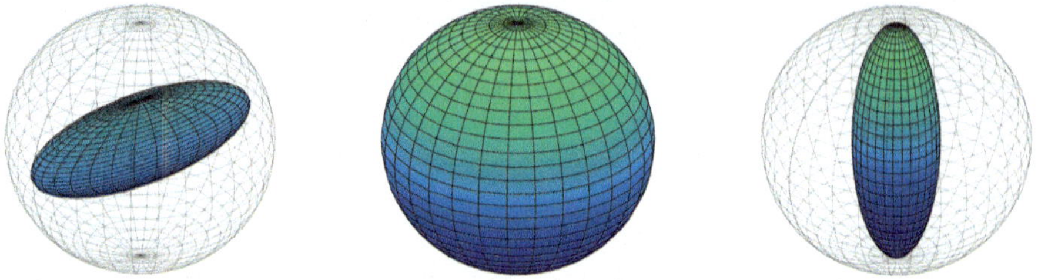

$\{X, \mathbb{1}\}$ and $p = 0.3$. $\{\mathbb{1}\}$, i.e., a noiseless evolution. $\{Z, \mathbb{1}\}$ and $p = 0.3$.

Fig. 4.1 Example of Pauli noise for 2-dimensional Hilbert space in Bloch sphere representation

A further refinement is achievable. In fact, from an information perspective, the global phase has no relevance. Hence, by considering the "quotient" group[1]

$$\mathbb{E} \equiv \{\mathbb{1}, \mathrm{X}, \mathrm{Z}, \mathrm{XZ}\}^{\otimes d} \subset \mathbb{P}, \tag{4.2}$$

one can assume $\{\mathrm{E}_k\}_k \subset \mathbb{E}$.

A comparison among Pauli noise evolution for a 2-dimensional state is given in Fig. 4.1.

4.2 Estimating an Evolution

For a given evolution $\mathcal{N}$, another useful representation of its action on a state σ is through its *Choi matrix* [5] $\varsigma_{\mathcal{N}}$. When the Kraus operators of $\mathcal{N}$ are known—i.e. $\{\sqrt{\mathrm{p}_k}\,\mathrm{E}_k\}_k$—the Choi matrix is immediately defined as [6]:

$$\varsigma_{\mathcal{N}} = \sum_{n,m} \sqrt{\mathrm{p}_n \mathrm{p}_m} |\mathrm{E}_n\rangle\rangle\langle\langle \mathrm{E}_m|, \tag{4.3}$$

with $|\mathrm{E}\rangle\rangle$ being the *vectorization* of E. When the evolution is *unitary*—i.e. $\mathcal{U}(\sigma) = \mathrm{U}\sigma\mathrm{U}^\dagger$—, Eq. (4.3) simplifies to

$$\varsigma_{\mathcal{U}} = |\mathrm{U}\rangle\rangle\langle\langle \mathrm{U}|. \tag{4.4}$$

Working with Choi matrices allows to compute the fidelity of some unknown evolution w.r.t. a target one. For example, given a target unitary evolution $\mathcal{U}$ and an *experimental* evolution $\mathcal{E}$, both operating on d-dimensional Hilbert space. The following function is a proper fidelity for the two channels:

[1] Being aware of the equivalence $\mathrm{XZ} \equiv -i\mathrm{Y}$.

$$\mathfrak{f}(\mathcal{U}, \mathcal{E}) = \frac{1}{d} \cdot \mathtt{Tr}(\sqrt{\sqrt{\varsigma_{\mathcal{U}}}\sqrt{\varsigma_{\mathcal{E}}}})^2. \tag{4.5}$$

Estimating $\varsigma_{\mathcal{E}}$ is an expensive procedure based on a orthogonal set of measurements. A sufficient and common one is the Pauli group. In the circuit model we refer to this procedure as follows

The circuit above represent a *state tomography*—i.e., how $\mathcal{E}$ affects σ. It works by running k sets of experiments to obtain an estimator for $\{\mathtt{p}_k\}_k$.

Similarly, a *process tomography* starts from an orthogonal set of input states in order to fully characterize $\varsigma_{\mathcal{E}}$. In the circuit model we express such an estimation as

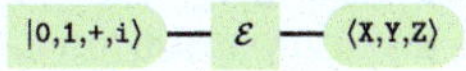

Tomography methods are intractable when considering high-dimensional systems.

4.3 Modeling Faulty Gates

According to the Pauli-Lindblad master equation [4, 7–9], a faulty operation can be modeled as $\mathcal{U} \circ \mathcal{N}$. In words, the model allows to think of a faulty operator as the composition of an ideal operator $\mathcal{U}$ preceded by some Pauli-noise $\mathcal{N}$—see Fig. 4.2.

For numerical evaluations, a common assumption is that a quantum evolution undergoes a depolarization [10] $\mathcal{D}$, which can be expressed as a mixture of Pauli errors:

$$\mathcal{D}(\sigma) = \sqrt{1 - \frac{3\lambda}{4}} \mathbb{1}\sigma\mathbb{1} + \sqrt{\frac{\lambda}{4}}(\mathtt{X}\sigma\mathtt{X} + \mathtt{Z}\sigma\mathtt{Z} + \mathtt{Y}\sigma\mathtt{Y}). \tag{4.6}$$

Starting from Eq. (4.6), it is possible to relate λ to a triplet of probabilities for the Pauli errors $\mathtt{X}$, $\mathtt{Y}$ and $\mathtt{Z}$. A method to do that is outlined in Ref. [11]. Furthermore, according to Ref. [2], the approximation

$$\mathcal{D}(\sigma) \approx (1 - \mathtt{p})^2\sigma + \mathtt{p}(1 - \mathtt{p})(\mathtt{X}\sigma\mathtt{X} + \mathtt{Z}\sigma\mathtt{Z}) + \mathtt{p}^2\mathtt{XZ}\sigma\mathtt{XZ}. \tag{4.7}$$

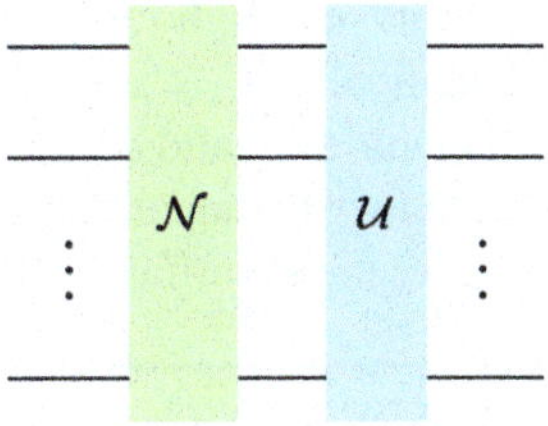

Fig. 4.2 Faulty operation $\mathcal{U}$ expressed as the composition $\mathcal{U} \circ \mathcal{N}$

Fig. 4.3 Propagation over $\wedge(\mathtt{X})$ operator

Fig. 4.4 Pauli noise affecting orthogonal single-qubit operators

is a good model for decoherence and it applies independently to each qubit of the system—i.e., $\mathcal{D}^{\otimes d}$, for a d-dimensional system.

The joint result of equation (4.7) together with the Pauli-Lindblad assumption, allows to investigate circuit as a composition of ideal operators affected by some single qubit Pauli error that affect the logic of computation. Such errors propagate among the circuit coherently with the logical operators. As basic example, consider the operator $\wedge(\mathtt{X})$. According to Eq. (4.7), a Pauli error may precede its execution and propagates through the system as in Fig. 4.3.

Despite the simplicity of the example, the two propagation rules shown in Fig. 4.3 are complete, as $\wedge(\mathtt{X})$ is the only non-single qubit necessary for universal computation.[2]

As regards single-qubit operators, consider operators $\mathtt{Z}^{\frac{1}{2}}, \mathtt{X}^{\frac{1}{2}}$, necessary to generate the Clifford group as Sect. 2.3. In such a case, orthogonal Pauli errors invert their logic, as in circuits of Fig. 4.4.

4.4 Error Correction and Logical Computing

4.4.1 Code Functions

Similarly to what is done in classical information [12], quantum information can be protected through the introduction of an *error correction* scheme [3]. This starts with the definition of a *code* function Γ, which introduce redundancy to the system and exploit it to restore the original information, in case an error occurs. Formally, a code function $\Gamma_{k,m} : \mathbb{H}^{\otimes k} \to \mathbb{H}^{\otimes(k+m)}$, able to encode k qubits into $k+m$ with $m > 0$. The code is said to have ratio $\frac{k}{k+m}$. When defining a code, a *desideratum* is to achieve a high ratio. Another likewise important metric is the *distance* of a code. Specifically, $\Gamma_{k,m}$ creates a *codeword* space s.t. $|\mathtt{Im}(\Gamma_{k,m})| = 2^{k+m}$ where only 2^k elements are valid codewords. Generally speaking a highly sparse valid space allows for higher distance among valid codewords. On contrary a dense valid space makes fuzzier the identification of a couple of valid-invalid codewords. This issue is treated

[2] We discussed this in detail in Sect. 2.

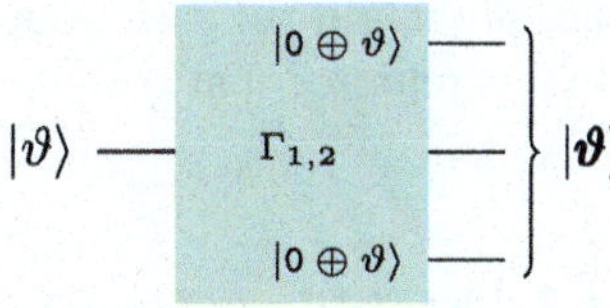

Fig. 4.5 Example of code function $\Gamma_{1,2} : \mathbb{H} \rightarrow \mathbb{H}^{\otimes 3}$

more formally in Sect. 4.7, while now it is just important to observe that distance and ratio are, by their nature, inversely proportional, creating a challenging trade-off to handle.

Redundancy Through Entanglement

As an example, consider a code function which takes in input a generic single qubit $|\vartheta\rangle = \alpha\,|0\rangle + \beta\,|1\rangle$ and generates a logical qubit

$$|\boldsymbol{\vartheta}\rangle \equiv \alpha\,|\mathbf{0}\rangle + \beta\,|\mathbf{1}\rangle \equiv \alpha\,|000\rangle + \beta\,|111\rangle\,. \tag{4.8}$$

Such code can be implemented as in Fig. 4.5.

Coherently with the definition of code, the system has 3 qubits with $2^k = 2$ possible outcomes: $|000\rangle$ and $|111\rangle$. After the application of $\Gamma_{1,2}$, some errors become detectable. For example, assume that the bit-flip noise $\mathcal{X} \mapsto \{\mathbb{1}, \mathtt{X}\}$ affects each qubit, with i.i.d. error probability $\mathtt{p}$. One can perform a projection over the even space and the odd space for qubits 1 and 2, and then the same for qubits 2 and 3. This can be done by performing the *non-destructive measurement*[3]

$$\mathtt{Z}^{\otimes 2} = (|00\rangle\langle 00| + |11\rangle\langle 11|) - (|01\rangle\langle 01| + |10\rangle\langle 10|). \tag{4.9}$$

The eigenvalues are ± 1 and the detectable errors are represented in the table below.

Syndrome	Detection
$+1, +1$	$\mathtt{I} \otimes \mathtt{I} \otimes \mathtt{I}$
$-1, +1$	$\mathtt{X} \otimes \mathtt{I} \otimes \mathtt{I}$
$-1, -1$	$\mathtt{I} \otimes \mathtt{X} \otimes \mathtt{I}$
$+1, -1$	$\mathtt{I} \otimes \mathtt{I} \otimes \mathtt{X}$

Notice that $\mathtt{Z}^{\otimes 2}$ has no effect on $\alpha\,|000\rangle + \beta\,|111\rangle$, whatever the target qubits are. For this reason its measurement can be used to detect some error, without affecting the original state.

This error correction scheme works for $\mathtt{p} < \frac{1}{2}$. Without the scheme the minimum fidelity is

$$\mathfrak{f} = \min_{\forall|\vartheta\rangle} \sqrt{\langle\vartheta|\,\mathcal{X}(|\vartheta\rangle\langle\vartheta|)\,|\vartheta\rangle} = \sqrt{1 - \mathtt{p}}.$$

[3] A possible implementation of non-destructive measurement is given in Sect. 4.6.

The probability of getting an error on the scheme is given by $\mathtt{p}_e = 3\mathtt{p}^2(1-\mathtt{p}) + \mathtt{p}^3$, so it is required that $\sqrt{1-\mathtt{p}_e} > \sqrt{1-\mathtt{p}}$, which happens when $\mathtt{p} < \frac{1}{2}$.

4.5 Stabilizer Codes

Generally speaking, defining an efficient code is a hard task. Here, we review a fundamental family called *stabilizer codes*, which is particularly helpful as it can relate to linear codes coming from classical error correction.

A code function $\Gamma_{k,m}$ is a stabilizer code if its image is characterized by an abelian subgroup[4] $\mathbb{S} \subseteq \mathbb{E}$ as follows

$$\mathtt{Im}(\Gamma_{k,m}) \equiv \{|\boldsymbol{\vartheta}\rangle \ : \ \mathtt{S}\,|\boldsymbol{\vartheta}\rangle = |\boldsymbol{\vartheta}\rangle\,, \forall \mathtt{S} \in \mathbb{S}\}$$

$\mathbb{S}$ must be abelian in order to stabilize a non-trivial code. Consider the case $[\mathtt{S}_1, \mathtt{S}_2] = 1$, then $\mathtt{S}_1\mathtt{S}_2\,|\boldsymbol{\vartheta}\rangle = -\mathtt{S}_2\mathtt{S}_1\,|\boldsymbol{\vartheta}\rangle = -\,|\boldsymbol{\vartheta}\rangle$, but also $\mathtt{S}_1\mathtt{S}_2\,|\boldsymbol{\vartheta}\rangle = |\boldsymbol{\vartheta}\rangle$; hence $|\boldsymbol{\vartheta}\rangle = -\,|\boldsymbol{\vartheta}\rangle = 0$ and $\mathbb{S}$ stabilizes the trivial code $\mathtt{Im}(\Gamma_{k,m}) = \{0\}$.

For 2^k possible outcomes, a stabilizer group $\mathbb{S}$ has 2^m elements and, since it is abelian, it can be specified by m generators $\{\mathtt{S}_1, \mathtt{S}_2, \ldots, \mathtt{S}_m\}$. The benefit of using generators is that to check whether a state vector is stabilized by $\mathbb{S}$ or not, one needs only to check it for the generators.

To see how the correction strategy works, consider an error operator $\mathtt{E} \in \mathbb{E}$. Let us analyse how $\mathtt{E}$ relates with a generator $\mathtt{S}_i$.

1. $\exists \mathtt{S}_i : \mathtt{E}\mathtt{S}_i = -\mathtt{S}_i\mathtt{E}$, then $\mathtt{S}_i\mathtt{E}\,|\boldsymbol{\vartheta}\rangle = -\mathtt{E}\mathtt{S}_i\,|\boldsymbol{\vartheta}\rangle = -\mathtt{E}\,|\boldsymbol{\vartheta}\rangle$. Therefore, $\mathtt{E}\,|\boldsymbol{\vartheta}\rangle$ is a -1 eigenvector of $\mathtt{S}_i$ and $\mathtt{E}$ can be detected by measuring $\mathtt{S}_i$.
2. Otherwise $\mathtt{E}\mathtt{S}_i = \mathtt{S}_i\mathtt{E}\ \forall \mathtt{S}_i$,[5] then if $\mathtt{E} \in \mathbb{S}$, it clearly doesn't corrupt the state. So the problem arises when $\mathtt{E} \notin \mathbb{S}$, making $\mathtt{E}$ *undetectable*.

The set of undetectable errors is given by $C_{\mathbb{E}}(\mathbb{S}) \setminus \mathbb{S}$, where C is the *centralizer* function. Nevertheless, a noise $\mathcal{N}$ with some undetectable operators, may still be *correctable*. Formally, a generic noise $\mathcal{N}(\sigma) = \sum_k p_k \mathtt{E}_k \sigma \mathtt{E}_k$ is correctable if any two operators $\mathtt{E}_i, \mathtt{E}_j \in \{\mathtt{E}_k\}_k$, differ in syndrome or have the same syndrome but differ by a stabilizer, i.e.

$$\mathtt{E}_i\mathtt{E}_j \in \mathbb{S} \cup \left(\mathbb{E} \setminus C_{\mathbb{E}}(\mathbb{S})\right).$$

[4] A group of which components commute one another. I.e. $[\mathtt{S}_i, \mathtt{S}_j] = 0$ holds for any $\mathtt{S}_i, \mathtt{S}_j \in \mathbb{S}$.

[5] Because any couple of $\mathbb{E}$ either commutes or anti-commutes.

4.6 Relation with Classical Binary Codes

Instead of expressing the error detection as the measurement operator $\mathtt{Z}^{\otimes 2}$, we can use the math coming from classical linear codes to define a quantum error correction scheme, e.g.

$$H = \begin{pmatrix} 1 & 1 & 0 \\ 0 & 1 & 1 \end{pmatrix}.$$

H can be used as parity-check matrix to detect and correct a bit flip occurring in one of the three physical qubits of $|\boldsymbol{\vartheta}\rangle$—see Fig. 4.6. Clearly, this is true as long as ancillary qubits undergoes a negligible noise.

Ancillary qubits allows to abstract from the hardware, while the real implementation of this scheme very depends on what kind of measurements the quantum processor supplies.[6] Since $\mathtt{E} \in \mathbb{E}$, one can represent a generic error as a $2(k+m)$-dimensional binary vector $(e_{\mathrm{x}}|e_{\mathrm{z}})$, such that if $e_{i,\mathrm{x}} = 1$, $\mathtt{E}$ has an $\mathtt{X}$ operator affecting the i-th qubit, or the identity otherwise. Symmetrically, for $e_{i,\mathrm{z}}$ the subject operator is $\mathtt{Z}$.

With the same criteria, let $(s_{i,\mathrm{x}}|s_{i,\mathrm{z}})$ be the binary vector representing a generator $\mathtt{S}_i$. Thus one can construct an $m \times 2(k+m)$ matrix H such that

$$H \equiv \begin{pmatrix} s_{1,\mathrm{x}}|s_{1,\mathrm{z}} \\ \vdots \\ s_{m,\mathrm{x}}|s_{m,\mathrm{z}} \end{pmatrix} \tag{4.10}$$

To check if the noise $\mathcal{N}$ is fully correctable, for any couple $(e_{i,\mathrm{x}}|e_{i,\mathrm{z}})$, $(e_{j,\mathrm{x}}|e_{j,\mathrm{z}})$, related to the operators $\mathtt{E}_i$, $\mathtt{E}_j$, the following holds:

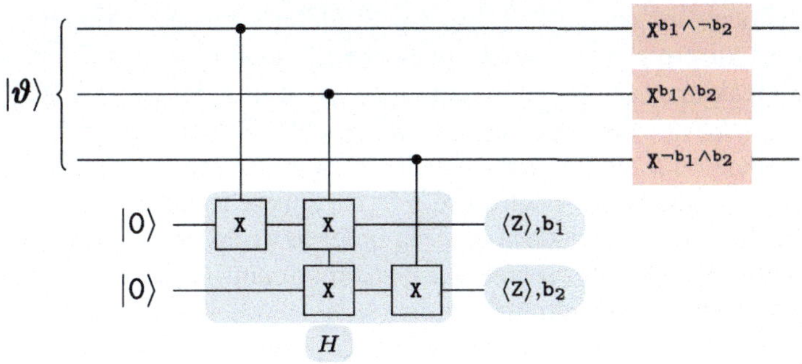

Fig. 4.6 The gray coloring helps to visualize how the classical matrix representation H of the detection scheme is implemented by means of auxilliary qubits

[6] See for example [13] for an experimental implementation, based on ancillary qubits to get non-destructive measurements.

$$H(e_{i,x} + e_{j,x} | e_{i,z} + e_{j,z})^{\mathsf{T}} \neq 0.$$

To get a stabilizer code starting from a classical linear code function[7] $\Sigma_{k,m} : \mathbb{F}_2^k \to \mathbb{F}_2^{(k+m)}$, with $0 < m < k$ and parity-check matrix H_Σ. The corresponding matrix for the quantum paradigm is given by

$$H_\Gamma \equiv \begin{pmatrix} H_\Sigma & \mathbf{0} \\ \mathbf{0} & H_\Sigma \end{pmatrix} \tag{4.11}$$

If the code is self-orthogonal—i.e. $\mathtt{Im}(\Sigma_{k,m})^\perp \subseteq \mathtt{Im}(\Sigma_{k,m})$—, H_Γ relates to a stabilizer group $\mathbb{S}$—in accordance with matrix (4.10)—for $k - m$ logical qubits and $k + m$ physical qubits.[8] Formally, following this procedure leads to a quantum code $\Gamma_{k-m,m}$ such that

$$\mathtt{Im}(\Gamma_{k-m,m}) \equiv \{|\boldsymbol{\vartheta}\rangle \ : \ \mathsf{S}\,|\boldsymbol{\vartheta}\rangle = |\boldsymbol{\vartheta}\rangle\,, \forall \mathsf{S} \in \mathbb{S}\}.$$

A slightly more general definition considers a couple of classical linear codes Σ_{k_1,m_1}, Σ_{k_2,m_2}. As long as $\mathtt{Im}(\Sigma_{k_2,m_2})^\perp \subseteq \mathtt{Im}(\Sigma_{k_1,m_1})$ and $k_1 + m_1 = k_2 + m_2$ hold, we can perform the parity check, in accordance with matrix (4.11). The result is a quantum code $\Gamma_{k,m}$ with $k = k_1 + k_2 - m_1 - m_2$ and $m = m_1 + m_2$. $\Gamma_{k,m}$ is called CSS code because of their creators Calderbank, Shor and Steane [14, 15].

4.7 Distance and Bounds

4.7.1 Classical Bounds

Consider a codeword set s.t. $|\mathtt{Im}(\Sigma_{k,m})| = 2^k$, defined in a 2^{k+m}-dimensional Hamming space. An error relates to a codeword u, creating a new word $\tilde{u} = u + e$. Generally speaking $\tilde{u}$ can relate to more than one u or e, making the definition of a good code a hard task. A good code should be able to relate any error e to one and one only codeword u. In this sense, such a code relates to each u a lattice sphere, centered in u with radius r. Each word $\tilde{u}$ in the sphere is such that $d(u, \tilde{u}) \leq r$ and $d(v, \tilde{u}) > r$ for any other codeword v. In order to avoid any overlap, the radius is upper-bounded by $r \leq \lfloor d/2 \rfloor$, where d is the minimum distance of the code. From this the *Hamming bound* follows:

$$\sum_{i=0}^{\lfloor d/2 \rfloor} \binom{k+m}{i} \leq 2^m$$

[7] $\mathbb{F}_2^n$ is a binary n-dimensional Hamming space, i.e. a vector of n binary values.

[8] Notice the loss in the ratio—i.e. $m/(k+m)$—caused by "quantizing" a classical code.

where 2^m is the maximum number of words orthogonal to the code and $\binom{k+m}{i}$ is the number of errors involving i bits. Whenever the spheres saturate this bound, without creating any overlap, the code is said to be *perfect*.[9]

The minimum possible distance to not run into overlaps is 3. To see that consider the basic case of two codewords `01`, and `11`. A parity-check with even result on the first codeword is indistinguishable from an odd result on the second codeword. It is possible to define a group of perfect codes[10] of distance $d = 3$. Namely, by setting $k = 2^h - h - 1$ and $m = h$ it results

$$\sum_{i=0}^{1} \binom{2^h - 1}{i} = 2^h$$

Notice how the ratio—i.e. $1 - \frac{h}{2^h-1}$—rapidly grows to 1 as h grows. More in general, for any high-dimensional code $\Sigma_{k,m}$ and minimum distance $d \propto k + m$, it is possible to prove [3] that the ratio $1 - \mathfrak{h}_2(d/(k+m))$ is achievable , where $\mathfrak{h}_2$ is the binary entropy.

4.7.2 *Quantum Bounds*

For a given noise $\mathcal{N}(\sigma) = \sum_k p_k \mathrm{E}_k \sigma \mathrm{E}_k$, let $\{\mathrm{E}_{k_i}\}_i \subseteq \{\mathrm{E}_k\}_k$ be the set of undetectable errors. Then, the code distance is given by

$$d = \mathtt{min}\big\{\mathfrak{w}(\mathrm{E}) \ : \ \mathrm{E} \in \{\mathrm{E}_{k_i}\}_i\big\},$$

where $\mathfrak{w}$ is the weight function, counting the number of single-qubit components differing from the identity operator $\mathtt{I}$. If the code $\mathtt{Im}(\Gamma_{k,m})$ is characterized by a stabilizer group $\mathbb{S}$, then

$$d = \mathtt{min}\big\{\mathfrak{w}(\mathrm{E}) \ : \ \mathrm{E} \in C_{\mathbb{E}}(\mathbb{S}) \setminus \mathbb{S}\big\}. \tag{4.12}$$

Symmetrically to classical codes, the Hamming bound related to a code $\Gamma_{k,m}$ is given by

$$\sum_{i=0}^{\lfloor d/2 \rfloor} 3^i \binom{k+m}{i} \leq 2^m.$$

The new factor 3^i expresses the possible error combinations from $\mathbb{E}$ involving any i qubits. Saturating the Hamming bound establishes a perfect code only if this is non-degenerate.

---

[9] Despite its name, the code is still unable to apply correction whenever $u_1 + e_1 = u_2 + e_2$ occurs, with u_1, u_2 being codewords and e_1, e_2 being errors.

[10] Known as Hamming codes.

Consider, as an example, the following stabilizer set

$$\mathbb{S} = \langle \mathtt{X}_{1,4}\mathtt{Z}_{2,3}, \mathtt{X}_{2,5}\mathtt{Z}_{3,4}, \mathtt{X}_{1,3}\mathtt{Z}_{4,5}, \mathtt{X}_{2,4}\mathtt{Z}_{1,5} \rangle$$

$\mathbb{S}$ generates a code $\Gamma_{k,m}$ such that $m = \log |\mathbb{S}| = 4$ and, as the operators are defined over a 5-dimensional system, $k = 1$. Ultimately, according to (4.12), the distance is given by any $\mathtt{E}$ operating on at least 3 qubits, coming from the centralizer group $C_{\mathbb{E}}(\mathbb{S})$, e.g. $\mathtt{E} = \mathtt{Z}_{1,2,4}\mathtt{X}_4$. To see that, consider any 1- or 2-qubits operator and check that it anti-commutes with some element from $\mathbb{S}$. Hence $d = \mathfrak{w}(\mathtt{E}) = 3$. The sphere coverage is

$$\sum_{i=0}^{1} 3^i \binom{5}{i} = 16 = 2^m$$

and, therefore, the code is perfect.

4.8 The Role of Stabilizers in Computing

There is a tight relation between communication and computing scenarios, as regards error correction. Namely, in communication, noise affects information conveyed through a physical channel. Similarly, in computing, noise affects information during the life-time of an algorithm. Both scenarios run under the same noise model $\mathcal{N}(\sigma) = \sum_k p_k \mathtt{E}_k \sigma \mathtt{E}_k$. However, during computation, logical states demand for the definition of *logical operators*.

A unitary operator $\mathbf{U}$ is a logical operator for a code Γ stabilized by $\mathbb{S}$ if, for any logical state $|\boldsymbol{\vartheta}\rangle$ and any $\mathtt{S} \in \mathbb{S}$, the following holds:

$$\mathbf{U}\,|\boldsymbol{\vartheta}\rangle \in \mathtt{Im}(\Gamma), \quad \mathbf{U}\mathtt{S}\mathbf{U}^\dagger \in \mathbb{S}.$$

To prove that, it is sufficient to show that

$$\mathbf{U}\mathtt{S}\mathbf{U}^\dagger\mathbf{U}\,|\boldsymbol{\vartheta}\rangle = \mathbf{U}\mathtt{S}\,|\boldsymbol{\vartheta}\rangle = \mathbf{U}\,|\boldsymbol{\vartheta}\rangle$$

holds for all $\mathtt{S}$ in the generator of $\mathbb{S}$. Notice that $\mathbf{U}\mathtt{S}\mathbf{U}^\dagger$ stabilizes $\mathbf{U}\,|\boldsymbol{\vartheta}\rangle$.

Defining a Stabilizer Code From Scratch

From the above emerges a general way to build a stabilizer code by defining together a code function $\bar{\Gamma}_{k,m}$ and its stabilizer group $\bar{\mathbb{S}}$. Formally consider the state $|\boldsymbol{\vartheta}\rangle \equiv |\vartheta\rangle \otimes |0\rangle^{\otimes m}$, which is (trivially) stabilized by the group $\mathbb{S} \equiv \langle \mathtt{Z}_{k+1}, \mathtt{Z}_{k+2}, \ldots, \mathtt{Z}_{k+m} \rangle$. Let $\mathtt{U}$ be a unitary operator mapping $\mathbb{S}$ to itself,[11] then

[11] This may be any operator coming from the Clifford group, as it satisfies the closure over the Pauli group.

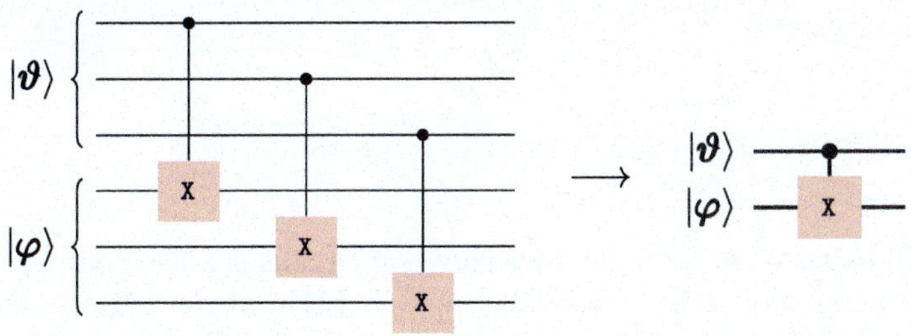

Fig. 4.7 Transversal implementation of a logical $\wedge(\mathbf{X})$

$$\mathtt{Im}(\bar{\Gamma}_{k,m}) = \{\mathtt{U}\,|\boldsymbol{\vartheta}\rangle \;:\; \bar{\mathtt{S}}\mathtt{U}\,|\boldsymbol{\vartheta}\rangle = \mathtt{U}\,|\boldsymbol{\vartheta}\rangle\,, \forall \bar{\mathtt{S}} \in \bar{\mathbb{S}}\},$$

where $\bar{\mathbb{S}} \equiv \langle \bar{\mathtt{S}}_1, \bar{\mathtt{S}}_2, \ldots, \bar{\mathtt{S}}_m\rangle$ and $\bar{\mathtt{S}}_i \equiv \mathtt{U}\mathtt{Z}_{k+i}\mathtt{U}^\dagger$.

If the code is meant for computation, the only missing ingredient is the set of logical operators, which is

$$\left\{\mathbf{Z}_i, \mathbf{X}_i \;:\; \mathbf{Z}_i \equiv \mathtt{U}\mathtt{Z}_i\mathtt{U}^\dagger,\; \mathbf{X}_i \equiv \mathtt{U}\mathtt{X}_i\mathtt{U}^\dagger\right\}_{1\le i\le k}.$$

Achieving Fault-Tolerant Computing

Let Γ be any stabilizer code satisfying *self-duality*[12] and being *doubly-even.*[13] Then the Clifford group generators $\wedge(\mathtt{X})$, $\mathtt{X}^{\frac{1}{2}}$ and $\mathtt{Z}^{\frac{1}{2}}$ relate to *fault-tolerant* logical operators $\wedge(\mathbf{X}), \mathbf{X}^{\frac{1}{2}}, \mathbf{Z}^{\frac{1}{2}}$.

These operators can be claimed to be fault-tolerant because they admit (in principle) a so-called *transversal* implementation, which is very efficient in terms of circuit depth and error propagation. Figure 4.7 shows an example of transversal implementation of $\wedge(\mathbf{X})$ between two logical qubits $|\boldsymbol{\vartheta}\rangle$ and $|\boldsymbol{\varphi}\rangle$. Its efficiency is given by the fact the each physical operator acts on independent pairs of physical qubits. For the same reason, also the noise does not mix up among the physical qubits.

As regard fault-tolerance for the universal gate set $\mathbb{C}^+$—see Sect. 2.3—and especially for the non-Clifford operator $\mathtt{Z}^{\frac{1}{4}}$; there are some proposal for transversal implementation for the logical operator $\mathbf{Z}^{\frac{1}{4}}$, but these usually do not relate to any stabilizer group. Hence, in literature, two main branches of research emerged:

- circuit manipulation with the goal of minimizing $\mathtt{Z}^{\frac{1}{4}}$ occurrences [16, 17];
- design of $\mathbf{Z}^{\frac{1}{4}}$ by means of *injection* protocols [18–20].

A basic example of $\mathtt{Z}^{\frac{1}{4}}$ *injection* is shown in Fig. 4.8; this performs the injection by introducing one auxiliary qubit to the processor, prepared in the state

$$|\omega\rangle = \frac{1}{\sqrt{2}}(|0\rangle + e^{\frac{i\pi}{4}}\,|1\rangle). \tag{4.13}$$

[12] $\mathtt{Im}(\Gamma) = \mathtt{Im}(\Gamma)^{\perp}$..

[13] Any codeword has Hamming weight divisible by 4.

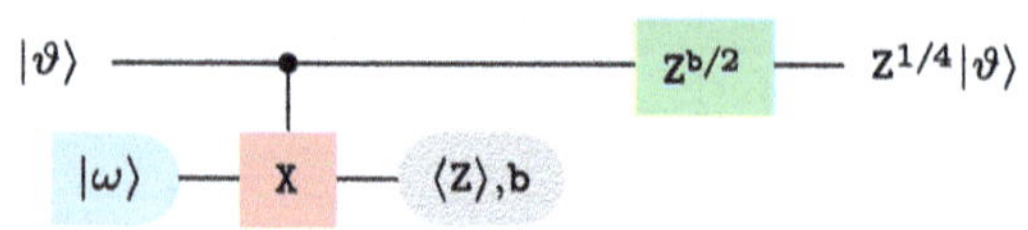

Fig. 4.8 Example of $Z^{\frac{1}{4}}$ gate *injection*

Normal forms—see Sect. 2.3—for universal circuits are also possible. An interesting result in this sense is available in Ref. [21], where authors showed that non-Clifford operators can be pushed to the beginning of the circuit.

4.9 Conclusion

With this chapter we covered many important topics to know when dealing with quantum computation. Especially considering the current state-of-the-art of quantum technologies, which are commonly referred as Noisy Intermediate-Scale Quantum (NISQ) architectures [22]. With the incoming Chap. 5, we will focus on optimizing circuits by means of *circuit compilation*, which preserves the circuit logic, while aiming to circuits more compliant to the hardware limitations. For practical reasons, we will consider circuit optimization without error correction schemes, as at the current stage of quantum technologies, these are to be considered at an early stage, where the gain promised by the theory struggle to be witnessed in real implementations. In fact, such schemes usually demands for more resources than the actual availability.

In accordance with our full-stack development proposed in Chap. 1, we expect error correction scheme implementations to show up at the *control system* level. Hence, a scheme will be in charge of providing a logical view of the physical resources. Because of such an organization, the compiler should not affect the logic on which the scheme relies on. This observation lead to think of new challenges specific to distributed architectures.

Before proceeding with the investigation of distributed compilers, we conclude the chapter by making some observation on the complication arising when trying and embedding error correction schemes within a distributed system.

4.9.1 Open Challenges

As already shown in Chap. 2, telegates work by means of the generation and distribution of Bell states. Implementing a *transversal logical telegate* would result into a high parallelism of each required task. However, it is not straightforward to *import* the quantum error correction schemes into the distributed paradigm. To understand why, consider Fig. 4.9, it shows all the telegate steps but the last one—i.e. the post-processing. The bold representation refer to a transversal implementation, according

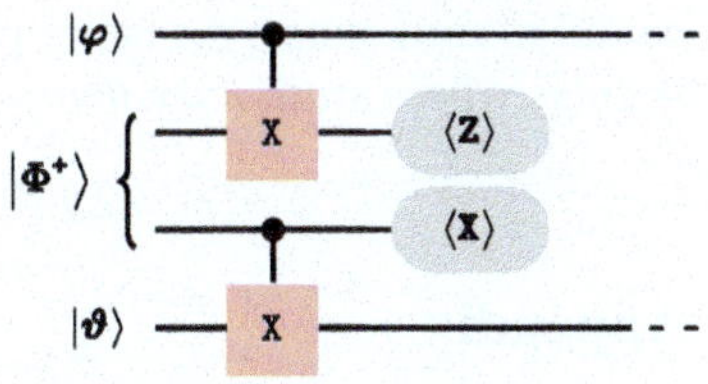

Fig. 4.9 Transversal telegate, without post-processing

to the formalism we introduced—see Fig. 4.7. Unfortunately, such an implementation is not logical in general, as the Bell states may affect the system which can end up *outside* the code scheme. For example, assuming:

$$\left|\boldsymbol{\Phi}^{+}\right\rangle \equiv |\Phi^{+}\rangle^{\otimes k+m} . \tag{4.14}$$

In such a case the Bell states are independent one another and because of this, there is a time lapse during which the system lays outside the code scheme and, for this reason, it is vulnerable to undetectable errors. The time lapse starts when applying the transversal operators $\wedge(\mathrm{X})$ and can only be restored by the post-processing, hoping that no error occurred in the meantime. In other words, the only detectable errors would be those caused by the post-processing, which, however, from a hardware perspective, results to be the most reliable step. Hence, this should not be considered as a practical way to proceed.

A possible way round is to generate and distribute a maximally entangled system, e.g., the generalized GHZ state:

$$\left|\boldsymbol{\Phi}^{+}\right\rangle \equiv \frac{1}{\sqrt{2}}\left(|0\rangle^{\otimes k+m} + |1\rangle^{\otimes k+m}\right). \tag{4.15}$$

Since the distributed system is now fully entangled, such a system can be used not only to perform the non-local operator $\wedge(\mathbf{X})$, but it results that the measurement outcome can be used to detect errors, combining so the syndrome with the post-processing.

The proposed distributed encoded system—Eq. (4.15)—seems to be the solution to the problem. However, it has significant drawbacks from an hardware perspective. It should be already clear, this far, that generating a Bell state with high fidelity and within a reasonable time lapse is very challenging. It naturally gets more complex, when thinking at higher degree systems, as the one of Eq. (4.15).[14]

In conclusion, to model a fault-tolerant scheme for $\wedge(\mathbf{X})$, Eq. (4.15) may be an assumption too strong to be practical. A more clever approach would be to define the generation and distribution protocol as to be **part of the encoding**. Very little has been done in this direction experimentally. An inspiring set-up can be found in Ref. [24], where authors managed to create a Shor code [25] by means of photons

[14] For example the smallest transversal code has $k = 1$ and $m = 6$ [23], resulting in the generation and distribution of a maximally entangled system of 14 qubits.

paired in Bell states. As being photon-based, this may bring to future experimental settings where *stationary-flying hybrid systems* are considered.

Appendix

4.10 Noise Canceling Through Indefinite Causal Orders

Similarly to what we have done in Sect. 2.7, we now report the possible advantages coming from the indefinite causal order framework, applied to non-unitaries. Namely, noisy channels are superposed in order to increase the overall capacity [26–29]

There are several ways to physically implement an indefinite order. Most of realizations are photonic-based [30–34], but it is not the only way. Indeed, within this Section we go over the work in Ref. [35]; presenting an implementation with a programmable technology, based on superconductors [36, 37].

The Indefinite Causal Orders for two evolution $\mathcal{N} \mapsto \{\mathrm{N}_n\}_n$ and $\mathcal{M} \mapsto \{\mathrm{M}_m\}_m$, is given by [38]:

$$\mathcal{S}(\sigma, \varkappa) = \sum_{nm} \mathrm{S}_{nm}(\sigma \otimes \varkappa)\mathrm{S}^{\dagger}_{nm}, \tag{4.16}$$

where $\varkappa$ is a *control state* and $\{\mathrm{S}_{nm}\}_{nm}$ denotes the set of Kraus operators such that

$$\mathrm{S}_{nm} = \mathrm{N}_n\mathrm{M}_m \otimes |0\rangle\langle 0| + \mathrm{M}_m\mathrm{N}_n \otimes |1\rangle\langle 1| \tag{4.17}$$

Therefore, under the assumption of errors in $\mathbb{E}$—see Eq. 4.2—, it is also true that

$$\mathrm{S}_{nm} = \sqrt{\mathrm{p}_n\mathrm{p}_m}\begin{bmatrix}\mathrm{E}_{nm} & \mathbf{0}\\ \mathbf{0} & \mathrm{E}_{mn}\end{bmatrix}. \tag{4.18}$$

As $\mathrm{N}_n\mathrm{M}_m = \sqrt{\mathrm{p}_n\mathrm{p}_m}\,\mathrm{E}_{nm}$ and $\mathrm{M}_m\mathrm{N}_n = \sqrt{\mathrm{p}_m\mathrm{p}_n}\,\mathrm{E}_{mn}$, with E_{nm}, E_{mn} being in the Pauli group.

As discussed in Sect. 2, higher-ordered circuit frameworks are not presented within this thesis. An *oracle* is sufficient to our purpose.

Syndrome	Detection
$+1, +1$	$\mathrm{I} \otimes \mathrm{I} \otimes \mathrm{I}$
$-1, +1$	$\mathrm{X} \otimes \mathrm{I} \otimes \mathrm{I}$
$-1, -1$	$\mathrm{I} \otimes \mathrm{X} \otimes \mathrm{I}$
$+1, -1$	$\mathrm{I} \otimes \mathrm{I} \otimes \mathrm{X}$

Our aim here is to report our work published as in [35]. Namely, experiencing and evaluating the indefinite causal order within a *Noisy Intermediate-Scale Quantum* (NISQ) architecture [22], based on superconductors. NISQ architectures are widespread and they promise to be resources of practical interest in the next future. Furthermore, the design is likely to rapidly evolve, also by considering the Indefinite Causal Order as resource. Our hope is to enrich the knowledge on the capabilities of current quantum technologies, with the long-term goal of contributing to shape future architecture designs.

The experiment set-up is meant to witness the communication advantage, resulting from a specific case of the subject evolution. According to *quantum Shannon theory* [39], the capacity is a metric to quantify the ability for a noisy channel to convey quantum information, without destroying it. A channel with null capacity destroys the *coherence* of the quantum information, meaning that it is not possible to retrieve the original information. By superposing two or more null capacity channels, the result is a new channel with a not-null capacity, an interesting behaviour from a practical point of view [27, 28].

In the following we consider the *bit-flip* channel $\mathcal{X} \mapsto \{\mathtt{X}, \mathbb{1}\}$ and the *phase-flip* channel $\mathcal{Z} \mapsto \{\mathtt{Z}, \mathbb{1}\}$, having noise probability, respectively, $\mathtt{p}$ and $\mathtt{q}$. Ultimately, by preparing the control state to be $\varkappa = |+\rangle\langle+|$, it occurs the evolution represented in Fig. 4.10 [27].

According to the bottleneck inequality [39], given the composite operation $\mathcal{Z} \circ \mathcal{X}$ and let $\mathfrak{C}(\cdot)$ be the quantum capacity, the following upper-bound holds [27]:

$$\mathfrak{C}(\mathcal{Z} \circ \mathcal{X}) \leq 1 - \mathtt{max}\{\mathfrak{h}_2(\mathtt{p}), \mathfrak{h}_2(\mathtt{q})\}, \tag{4.19}$$

where $\mathfrak{h}_2$ denotes the binary Shannon entropy. Also, the same inequality holds for $\mathcal{X} \circ \mathcal{Z}$.

Whenever both $\mathtt{p}$ and $\mathtt{q}$ are equal to $\frac{1}{2}$, we have that both the configurations are characterized by a null capacity, i.e., $\mathfrak{C}(\mathcal{Z} \circ \mathcal{X}) = \mathfrak{C}(\mathcal{X} \circ \mathcal{Z}) = 0$.

Let us now superpose the two noises. Accordingly, with probability $\mathtt{pq}$ the output of circuit 4.10 is given by the second addendum, namely, $\left(\mathtt{Y}\sigma\mathtt{Y}\right) \otimes |-\rangle\langle-|$. As consequence the capacity of $\mathcal{S}$ is lower-bounded by $\mathfrak{C}(\mathcal{S}) \geq \frac{1}{4}$, as shown in [28]. Specifically, σ occurs to pass through $\mathcal{Y}(\sigma) = \mathtt{Y}\sigma\mathtt{Y}$, coherently with control state being $|-\rangle\langle-|$. Therefore, it is possible to exploit the control state to gain a heralded unitary evolution $\mathcal{Y}$ via post-selection through the occurrence of $|-\rangle\langle-|$. Since $\mathcal{Y}$ is unitary, it is also reversible, therefore we can restore the information, gaining a perfect transmission of σ, i.e.,

$$\mathcal{Y} \circ \mathcal{Y}(\sigma) = \mathtt{YY}\sigma\mathtt{YY} = \sigma. \tag{4.20}$$

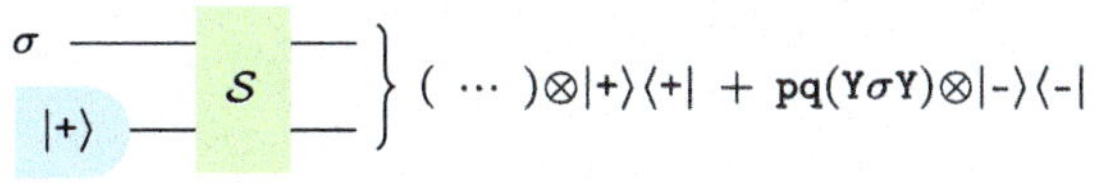

Fig. 4.10 Superposition of causal orders for bit-flip and phase-flip

4.10.1 Quantum Simulation

In this section we present our steps to realize $\mathcal{S}$ of Fig. 4.10. We already discussed that the state of the art on quantum technologies doesn't allow any native implementation of the indefinite causal orders. Hence our goal is to witness the communication advantage through quantum simulation.

When considering non-unitaries, as in the case of our interest, the overhead grows up w.r.t. superposing unitaries. The reason is that non-unitaries are naturally harder to engineer and need to be simulated as well. Specifically, for any Hilbert space $\mathbb{H}$, the corresponding set of density states lies in the convex map $\mathcal{C}(\mathbb{H})$.

According to the *Stinespring dilation* [40], one can always associate to a non-unitary evolution $\mathcal{N} : \mathcal{C}(\mathbb{H}) \to \mathcal{C}(\mathbb{H})$ a unitary one, $\mathcal{A}_{\mathcal{N}}$, defined as follows:

$$\mathcal{A}_{\mathcal{N}} : \mathcal{C}(\mathbb{H}) \otimes \mathcal{C}(\mathbb{A}) \to \mathcal{C}(\mathbb{H}) \otimes \mathcal{C}(\mathbb{A}) \tag{4.21}$$

where $\mathcal{C}(\mathbb{A})$ is an auxiliary system with associated basis $\left\{ |a_v\rangle \langle a_w| \right\}_{vw}$. Since $\mathcal{A}_{\mathcal{N}}$ is unitary, it has direct realization with the circuit algorithm. To simulate $\mathcal{N}$ from a realization of $\mathcal{A}_{\mathcal{N}}$, one need to *discard* the auxiliary system. In terms of operations, discarding the auxiliary system means applying a *partial trace* $\mathtt{Tr}_2 : \mathcal{C}(\mathbb{H}) \otimes \mathcal{C}(\mathbb{A}) \to \mathcal{C}(\mathbb{H})$. Specifically, for a generic state $\sigma = \sum_{ijvw} c_{ijvw} \left(|\vartheta_i\rangle \langle\vartheta_j| \otimes |a_v\rangle \langle a_w| \right)$, the partial trace outputs the following [41]:

$$\mathtt{Tr}_2(\sigma) = \sum_{ijvw} c_{ijvw} \, |\vartheta_i\rangle \langle\vartheta_j| \, \langle a_w | a_v\rangle. \tag{4.22}$$

Since $\mathbb{H}$ and $\mathbb{A}$ are taken to be generic systems, Eq. (4.22) has a direct generalization to the form $\mathtt{Tr}_{i_1,\ldots,i_k}$, tracing out subsystems indexed by $i_1, \ldots, i_k$.

In summary, we just outlined a method to realize an operation $\mathcal{N}$, involving two steps:

1. realizing the circuit $\mathcal{A}_{\mathcal{N}}$;
2. discarding the auxiliary system with a partial trace $\mathtt{Tr}_2$.

To our purpose, we apply this method, restricted to evolution $\{\mathtt{E}, \mathbb{1}\}$. In circuit representation this is shown in Fig. 4.11. To superpose them we need an extra qubit, which encodes the control system. The final realization is shown in Fig. 4.12.

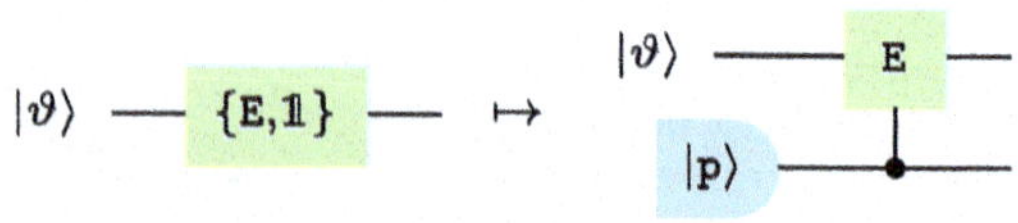

Fig. 4.11 Stinespring dilation simulating a generic Pauli noise E

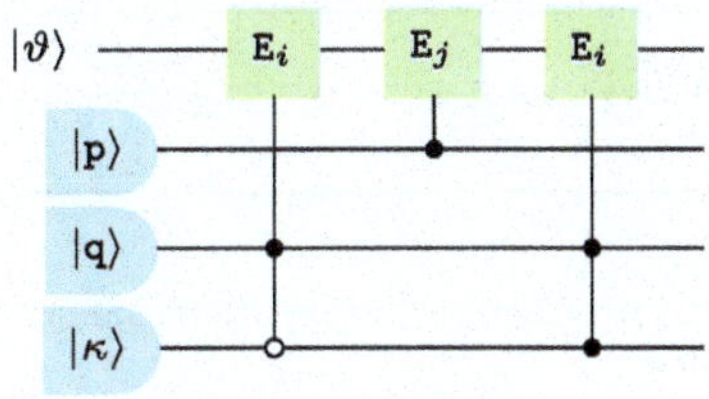

Fig. 4.12 Circuit implementing an indefinite causal orders between two Pauli noise $\mathtt{E}_i$, $\mathtt{E}_j$

4.10.2 Physical Setting

Starting from circuit 4.12, we can do a step closer towards the real physical setting meant for us to witness the communication advantage of the indefinite causal order. To this aim we need to simulate the evolution from Fig. 4.10, for the case of i.i.d. probabilities—i.e. $p = q = \frac{1}{2}$. Figure 4.13 shows the physical setting we used for our experiments. Notice that we added specific state preparations and measurements for information and control qubits. This update express a process tomography settings, explained in detail in Sect. 4.2. Second and third qubits are not measured, rather, their final output is ignored, which naturally express trace out over those systems—i.e. $\mathtt{Tr}_{2,3}$.

At the end each run a post-selection occurs. Namely, coherently with our discussion of Sect. 4.10, we only keep those outputs where the control qubit results in the state $|-\rangle$. To the given set of outputs, we then apply a **classical bit-flip** in case the information qubit was subject to a measurement $\langle \mathtt{X} \rangle$ or $\langle \mathtt{Z} \rangle$. In fact, aware of the fact that the communication advantage comes from the post-processing of equation (4.20), which is a Pauli operation, this can be computed classically and has an effect only when measuring $\langle \mathtt{X} \rangle$ and $\langle \mathtt{Z} \rangle$.

Circuit Decomposition and Optimization

As already discussed in Sect. 2.7.2, it is often the case that a real quantum technology doesn't supply natively a gate. For our setting of Fig. 4.13, this is the case of the 3-qubits gates, which have expensive decomposition [42].

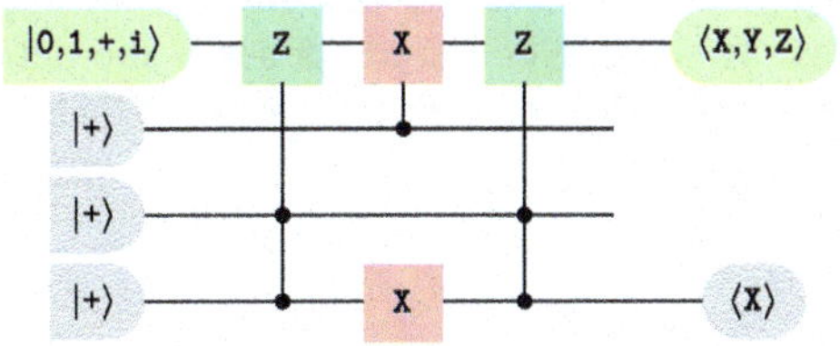

Fig. 4.13 Circuit representation of the quantum experiment setting we used to witness the communication advantage

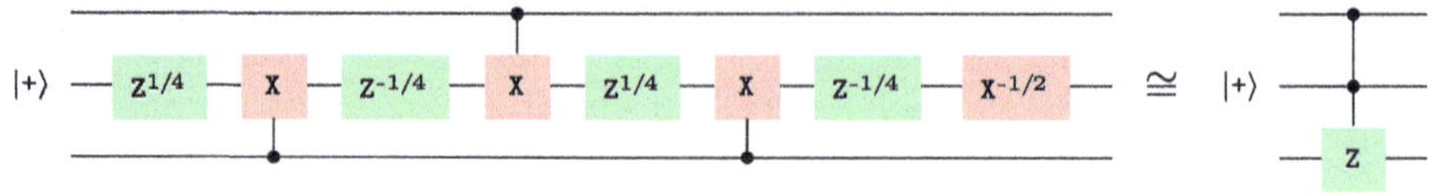

Fig. 4.14 Whenever the second wire takes $|+\rangle$ as input, the gate r.h.s. is equivalent, up to a global phase, to the decomposition l.h.s

Fig. 4.15 Bloch sphere representation of the experimental characterization for the physical setting of Fig. 4.13

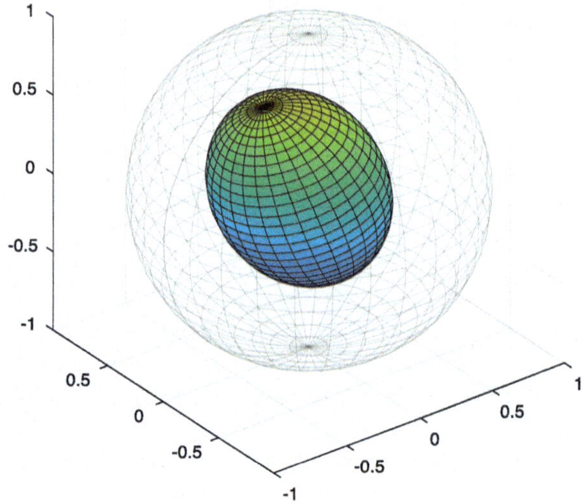

To witness the communication advantage we managed to do some optimization on the first occurrence of such a gate. The rationale behind the optimization is that the gate decomposition could be simpler in case we have some knowledge of the input. We indeed know there is at least one qubit prepared in the state $|+\rangle$. This is enough to use the decomposition in Fig. 4.14 instead of the standard one.

The final result is shown by plotting the characterization of the channel as a bloch sphere—See Fig. 4.15. It represents the experimental characterization for the physical setting of Fig. 4.13. Gray sphere represents the ideal sphere, corresponding to a set of pure states. The inside coloured sphere is the deformation induced by the imperfections caused by the employed technology, which in this case is the `santiago` processor provided by `IBM`,[15] which has a *quantum volume* [43] of 16.

[15] The processor has been retired at the time of writing.

References

1. Nielsen MA, Chuang IL (2010) Quantum computation and quantum information. Cambridge University Press
2. Dennis E et al (2002) Topological quantum memory. J Math Phys 43(9):4452–4505
3. Mancini S, Winter A (2020) A quantum leap in information theory. World Scientific
4. van den Berg E et al (2002) Probabilistic error cancellation with sparse Pauli-Lindblad models on noisy quantum processors. arXiv preprint arXiv:2201.09866
5. Choi M-D (1975) Completely positive linear maps on complex matrices. Linear Algebra Appl 10(3):285–290
6. de Jong J (2019) Fault-tolerant quantum computation: implementation of a fault-tolerant SWAP operation on the IBM 5-qubit device. MA thesis. Delft University of Technology
7. Breuer H-P, Petruccione F et al (2002) The theory of open quantum systems. Oxford University Press on Demand
8. Schlosshauer MA (2007) Decoherence: and the quantum-to-classical transition. Springer Science & Business Media
9. Cacciapuoti AS, Caleffi M (2019) Toward the quantum internet: a directional-dependent noise model for quantum signal processing. In: ICASSP 2019-2019 IEEE international conference on acoustics, speech and signal processing (ICASSP). IEEE, pp 7978–7982
10. Nielsen MA (2003) Quantum computation by measurement and quantum memory. Phys Lett A 308(2–3):96–100
11. Gidney C. Decorrelated depolarization. `https://algassert.com/post/2001`
12. Baylis DJ (2018) Error correcting codes: a mathematical introduction. Routledge
13. Cramer J et al (2016) Repeated quantum error correction on a continuously encoded qubit by real-time feedback. Nat Commun 7(1):1–7
14. Calderbank R, Shor P (1996) Good quantum error-correcting codes exist. Phys Rev A 54(2):1098
15. Steane A (1996) Multiple-particle interference and quantum error correction. Proc R Soc Lond. Ser A: Math, Phys Engin Sci 452(1954):2551–2577
16. Selinger P (2013) Quantum circuits of T-depth one. Phys Rev A 87(4):042302
17. Amy M, Maslov D, Mosca M (2014) Polynomial-time T-depth optimization of Clifford+T circuits via matroid partitioning. IEEE Trans Comput-Aided Design Integrat Circuits Syst 33(10):1476–1489
18. Yoganathan M, Jozsa R, Strelchuk S (2019) Quantum advantage of unitary Clifford circuits with magic state inputs. Proc R Soc A 475(2225):20180427
19. Li Ying (2015) A magic state's fidelity can be superior to the operations that created it. New J Phys 17(2):023037
20. Zhou X, Leung DW, Chuang IL (2000) Methodology for quantum logic gate construction. Phys Rev A 62(5):052316
21. van de Wetering J (2021) Constructing quantum circuits with global gates. New J Phys 23(4):043015
22. Preskill J (2018) Quantum Computing in the NISQ era and beyond. Quantum 2(79)
23. Albert VV, Faist P (eds) (2022) [[7, 1, 3]] Steane code. In: The error correction zoo. URL: https://errorcorrectionzoo.org/c/steane
24. Zhang R et al (2022) Loss-tolerant all-photonic quantum repeater with generalized Shor code. Optica 9(2):152–158
25. Albert VV, Faist P (eds) (2022) [[9, 1, 3]] Shor code. In: The error correction zoo. URL: https://errorcorrectionzoo.org/c/shor_nine
26. Ebler D, Salek S, Chiribella G (2018) Enhanced communication with the assistance of indefinite causal order. Phys Rev Lett 120(12):120502
27. Salek S, Ebler D, Chiribella G (2018) Quantum communication in a superposition of causal orders. arXiv preprint arXiv:1809.06655
28. Caleffi M, Cacciapuoti AS (2020) Quantum switch for the quantum internet: noiseless communications through noisy channels. IEEE J Select Areas Commun 38(3):575–588

29. Koudia S et al (2022) How deep the theory of quantum communications goes: Superadditivity, superactivation and causal activation. In: IEEE communications surveys and tutorials
30. Procopio LM, Moqanaki A, Araújo M et al (2015) Experimental superposition of orders of quantum gates. Nat Commun 6(1):1–6
31. Rubino G et al (2017) Experimental verification of an indefinite causal order. Sci Adv 3(3):e1602589
32. Guo Y et al (2020) Experimental transmission of quantum information using a superposition of causal orders. Phys Rev Lett 124(3):030502
33. Fitzsimons JF, Jones JA, Vedral V (2015) Quantum correlations which imply causation. Sci Rep 5(1):1–7
34. Goswami K et al (2018) Indefinite causal order in a quantum switch. Phys Rev Lett 121(9):090503
35. Cuomo D, Caleffi M, Cacciapuoti AS (2021) Experiencing the communication advantage of the Superposition of Causal Orders. In: IEEE 22nd international workshop on signal processing advances in wireless communications (SPAWC). IEEE, pp 181–185
36. IBM Quantum (2021) https://quantum-computing.ibm.com/
37. Castelvecchi D (2017) IBM's quantum cloud computer goes commercial. Nature 543(7644):159
38. Chiribella G et al (2013) Quantum computations without definite causal structure. Phys Rev A 88(2):022318
39. Wilde MM (2013) Quantum information theory. Cambridge University Press
40. Forrest Stinespring W (1955) Positive functions on C*-algebras. Proc Am Math Soc 6(2):211–216
41. Rieffel EG, Polak WH (2011) Quantum computing: a gentle introduction. MIT Press
42. Shende VV, Markov IL (2009) On the CNOT-cost of TOFFOLI gates. Quantum Inf Comput 9(5):461–486
43. Cross AW et al (2019) Validating quantum computers using randomized model circuits. Phys Rev A 100(3):032328

Chapter 5
Circuit Synthesis and Resource Optimization

As outlined in Chap. 1, a full-stack development of a distributed system for quantum computation requires to be carefully engineered. The proposed stack allows a circuit designer to focus on the logic of its algorithm, without necessary consider all the issues coming from the physical infrastructure that will take care of computing it.

In this chapter we consider the logical layers interfacing with the algorithm (written in circuit model). These take care of *optimizing the circuit*, adapting it to the constraints given by the physical layers. Historically, a module called *compiler* does all the job and the corresponding optimization problem is called the *compilation problem*. This is generally a tough task to solve, even on single processor, and for which an NP-hardness proof is available [1]. An ever growing literature arises with a variety of proposals for local computation [2–16] and for distributed computation [17–26].

Whilst quantum processors are already available, distributed architectures are at an early stage. For this reason we need to identify the main components of such technologies. A key component supplies *telegates* as the fundamental inter-processor operations [27–29]. We already discussed in Chap. 2 how telegates work, but we report below the main steps.

Each telegate can be decomposed into several tasks, that we group as follows: (i) the generation and distribution of entangled states among different processors, (ii) local operations and (iii) classical communications. These tasks makes the telegate an expensive resource, in terms of running time and/or fidelity. As a consequence, they have critical impact on the performance of the overall computation. In contrast to such a limit, telegates offer remarkable opportunities of parallelization. In fact, much circuit manipulation is possible to keep computation independent from telegate's tasks. Therefore, we aim to model an optimization problem that embeds such opportunities. To do so we propose *a role* for both compiler and scheduler, within the stack defined in Chap. 1. The compiler anticipates the scheduler, facilitating its job in *selecting network resources*.

D. Cuomo, *Architectures and Circuits for Distributed Quantum Computing*,
Springer Theses, https://doi.org/10.1007/978-3-031-73808-1_5

5.1 Compiler's Strategy

The compiler is responsible for the mapping of an *unstructured* quantum circuit into a *structured* one. With this procedure we obtain two advantages. One is the direct consequence of working with structured circuits: having knowledge of the circuit structure allows to give formal definition of the optimization problem. Furthermore, we can look for structures that are compliant with a distributed architecture.

A compiler is a map $\mathcal{C} \mapsto \mathcal{L}$, where $\mathcal{C}$ is an unstructured circuit and $\mathcal{L}$ is a layered circuit. A layer $\ell_t \in \mathcal{L}$ runs at time-step t. In order to do so, the scheduler will take care of selecting the entanglement links to be used.

As anticipated, the compiled circuit $\mathcal{L}$ features some properties that inform us about the kind of gates occurring at a certain time-step. More precisely, for any given time-step t, the compiler produces a layer ℓ_t, composed by a global gate—e.g. $\wedge(Z^{\otimes m})$.

We specify the compiler depending on what group the input algorithm belongs. We start by addressing the Clifford group and conclude with other relevant groups in the context of universal computation.

5.2 Scheduler's Strategy

The scheduler is responsible for the allocation of the resources used to run global gates. At each time-step, the scheduler generates a resource demand, selecting which entanglement links have to be established. For this reason, the scheduler holds a representation of the physical layer. For the sake of reasoning, we assume one computation qubits per processor. The communication qubits are fixed and their derivation comes from the network topology.

According to the protocols we defined in Chaps. 1 and 3, the interaction among qubits requires the use of some entanglement links. To connect those distant qubits, the scheduler outputs an entanglement tree. The search for a tree coverage can be expressed as a generalization of the *minimum spanning tree* problem [30, Chap. 4.5]. Such a generalization is known as *minumum Steiner tree* problem [31], which is not tractable. Nevertheless efficient approximation ratio have been achieved [32–34] and it can be used for any topology. A further interesting result is that for lattice topologies the problem admits a *polynomial-time approximation scheme* [35]. Finally, we report a result important to us within the following remark [36, 37]:

Remark The minimal Steiner tree on rectangular lattices can be found in polynomial time.

For this reason, we will work with rectangular lattices as representation of the physical layers.

5.3 Objective Function

To optimize a circuit, we need to identify the *objective function*, so that we give a rate to the expected performance. A common approach is to evaluate only those operators which are somehow a bottleneck to computation. Considering the universal gate set $\mathbb{C}^+$—defined in Sect. 2.3, in the context of fault-tolerant quantum computing [38], the bottleneck is the $\mathtt{Z}^{1/4}$ operator [39, 40], since error correction protocols are designed for the Clifford group $\mathbb{C}$. Conversely, on current NISQ technologies, the bottleneck lies in the interaction between qubits—as for the case of $\wedge(\mathtt{X})$. The relevant metric can either be the number of occurrences of some operator $\mathtt{O}$, namely the $\mathtt{O}$-count, or the number of layers containing $\mathtt{O}$ at least once, namely the $\mathtt{O}$-depth. To rate a compiled circuit on distributed architectures, we do something along the lines of this approach. Specifically, the bottleneck are the non-local operators, each of which implies one occurrence of entanglement generation and distribution stage. We refer to such a stage as the $\mathfrak{E}$ operator. Therefore, we will rate a circuit by means of its $\mathfrak{E}$-depth and $\mathfrak{E}$-count.

The above remark tells us that we can find, for any global gate, the minimum $\mathfrak{E}$-count possible, by constructing an entanglement tree treated as a minimum Steiner tree problem. Figure 5.1 shows an example computed with the method provided by the `networkx` library [41].

5.3.1 *Reduction to Normal Form*

Normal forms generally provide a useful perspective of something under study. Being able to rely on a well defined structure can be very helpful for the design of a complex framework as the one under analysis.

If we can express the Clifford group in some normal form, we may be able to define optimization strategies that can relax their assumption on the problem they are tackling. Potentially, normal forms can reduce and simplify the domain we have to consider. Clifford group is known to be a first useful tool to characterize quantum

(a) Link availability. (b) An entanglement tree.

Fig. 5.1 Example of entanglement-based architecture, with nine nodes disposed and connected as a rectangle lattice

Fig. 5.2 Any Clifford function can be built as a composition of 3 blocks [42]

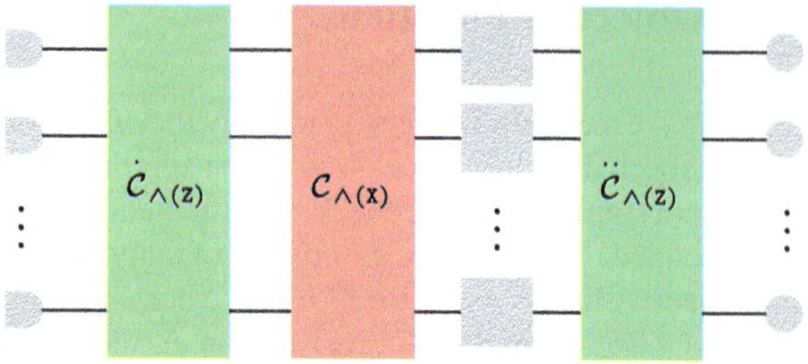

technologies. Even when expressed in normal form, its property holds.[1] For this reason, throughout this thesis we often refer to the Clifford group to benchmark our framework.

It has been shown in [42] that any Clifford function can be represented in the normal form depicted in Fig. 5.2. This normal form is of practical interest as it can be obtained starting from any Clifford circuit, which is in general not in normal form. Such a result comes from the employment of a ZX-*calculus* reasoner, e.g. [43]. ZX-calculus [42, 44] is a graphical language, arisen as an optimizer for quantum circuits, that translates a quantum circuit into a ZX-*diagram*. The main difference between the diagram and the original circuit is that the former works with ZX-*rules*, which serve as a reasoning tool to smartly generate a new circuit, equivalent to the original one. ZX-calculus was introduced in the literature in 2007 [45], with the main objective of minimizing a circuit gate-depth, and its potential is attracting the interest of researchers coming from different fields. In fact, we use it here to perform architecture-compliant optimization.

Coming back to Fig. 5.2, we use the circuit symbol — to express a generic Pauli state preparation. Similarly, the symbol — expresses a generic Pauli measurement.

The normal form suggests that the problem can be separated into three parts, corresponding to $\dot{\mathcal{C}}_{\wedge(\mathrm{Z})}$, $\mathcal{C}_{\wedge(\mathrm{X})}$ and $\ddot{\mathcal{C}}_{\wedge(\mathrm{Z})}$. For two of them—i.e., $\dot{\mathcal{C}}_{\wedge(\mathrm{Z})}$ and $\ddot{\mathcal{C}}_{\wedge(\mathrm{Z})}$—the order relation is trivial (as all $\wedge(\mathrm{Z})$ commute), which gives a lot of room for optimization. A bit trickier are instead $\mathcal{C}_{\wedge(\mathrm{X})}$ gates, as these do not commute in general. For this reason, we treat $\mathcal{C}_{\wedge(\mathrm{X})}$ and $\mathcal{C}_{\wedge(\mathrm{Z})}$ circuits differently.

With our previous work published in [46], we addressed the optimization of $\mathcal{C}_{\wedge(\mathrm{Z})}$ circuits. We here enhance our strategies and also include $\mathcal{C}_{\wedge(\mathrm{X})}$ circuits. We close the investigations with an extensive performance analysis of our framework for Clifford circuits. This will be followed by investigations for other groups of circuit, important for their role in universal computation.

[1] Clifford circuit can be used to characterize the noise model affecting an architecture, with a precision up to its standard deviation.

(a) Fully commuting circuit $\mathcal{C}_{\wedge(\mathrm{z})}$. (b) Non-commuting circuit $\mathcal{C}_{\wedge(\mathrm{x})}$, in general.

5.3.2 Layering $\mathcal{C}_{\wedge(\mathrm{z})}$ Circuits

Starting with $\mathcal{C}_{\wedge(\mathrm{Z})}$, these features two properties we can take advantage of:

1. $\mathcal{C}_{\wedge(\mathrm{Z})}$ is fully commuting;
2. $\wedge(\mathrm{Z})$ gates are symmetrical.

Because of (1), the compiler can re-arrange the gates in $\mathcal{C}_{\wedge(\mathrm{Z})}$ in any order, preserving the unitary function to be computed. For this reason, it is possible to *gather* any set of m gates that share the same control qubit into a global gate $\wedge(\mathrm{Z}^{\otimes m})$. Also, feature (2) means that control and target qubits are interchangeable. Hence we can use any qubit as *pivot* to create a global gate. The best we can do is to start from the qubit with the most number of interactions. This is the first pivot selected by the compiler. The first layer ℓ_1 is composed by a global gate with such a qubit as control. After that, we remove from the evaluation the gathered gates and proceed to build ℓ_2 the same way. We proceed iteratively until there are no more gates to compile. The result—i.e. $\mathcal{L}_{\wedge(\mathrm{Z})}$—is a sequence of global gates with decreasing number of targets.

5.3.3 Layering $\mathcal{C}_{\wedge(\mathrm{x})}$ Circuits

$\mathcal{C}_{\wedge(\mathrm{X})}$ computes a linear bijective parity function $g : \mathbb{F}_2^n \rightarrow \mathbb{F}_2^n$ [47]. Such a function is also defined by a n-by-n matrix M, with elements in the boolean space $\mathbb{F}_2$, iff

$$|g(\mathrm{b})\rangle = \mathrm{M}|\mathrm{b}\rangle, \tag{5.1}$$

for any basis string b.

For a given control-target system $|q_u, q_v\rangle$, a $\wedge(\mathrm{X})$ gate operating on it can be written as

$$\mathrm{R}_{uv} \equiv \mathbb{1} + r_{uv},$$

with r_{uv} being the zero matrix and one on element u, v.

The matrix M can be derived by $\mathcal{C}_{\wedge(\mathrm{X})}$ by multiplying its elements R_{uv}. Any decomposition M into R_{uv} matrices is a valid implementation of g. There are a number of methods generating different decompositions [47].

Once identified a matrix M, the compiler *gathers* rows to form global gates. Specifically, we can find, within the decomposition of M, sequences $R_{uv_1}, R_{uv_2}, \ldots, R_{uv_m}$ relating to global gates $\wedge(X^{\otimes m})$. This is a greedy process.

5.3.4 Analysis on the Upper-Bounds and Future Perspective

There is a fair doubt arising from the employment of normal forms for compilation: do we know the overhead cause by mapping any Clifford circuit to some normal form? If yes, is it reasonable?

The answer is positive to both questions. By working with normal forms, we are not only able to work with a circuit with known shape, but we can also upper-bound the overhead for the number of introduced operations. Depending on whether or not ancillae are considered, the system get more complex in terms of space or run-time. In Ref. [48], authors treat both cases, and prove linear upper-bounds.

Normal forms unlock also better opportunities from an hardware perspective. Specifically, dealing with well defined circuit allows to extend the gate set with more practical operators, as the $\wedge(X^{\otimes m})$ introduced in Chap. 2. From a hardware perspective, these are also commonly referred as to act *globally* and *simultaneously* [49–52]. Citing [49]: "It has been suggested that polynomial or exponential speedups can be obtained with global [gates]".

Other results in terms of overhead can be found in Ref. [53], where authors proved that any n-qubit Clifford circuit can be synthesised to $4n - 6$ global gates and any n-qubit circuit with $\dot{n}$ non-Clifford gates can be synthesised with no more than $2\dot{n} + \mathcal{O}(n/\log n)$ global gates.

Ultimately, our choice to employ the normal form of Fig. 5.2 has several benefits, besides the ones we already discussed:

- It is practical, as the open-source `pyzx` [43] provides the tools to perform the mapping.
- It is efficient, as the `pyzx` engine works to minimize the number of two-qubit gates.
- It has a good shape, as $\mathcal{C}_{\wedge(Z)}$ circuits are generally easier to optimize than $\mathcal{C}_{\wedge(X)}$ ones.

5.3.5 Clifford Performance Analysis

We conclude the evaluation of our framework with this section. We now tackle the compilation on Clifford circuits uniformly generated at random.

On behalf of techniques and results we introduced within this thesis, we now fix some parameters, such as, graph topology and compilation strategy. Our interest now is to look out at the performances a distributed architecture would achieve. We argued

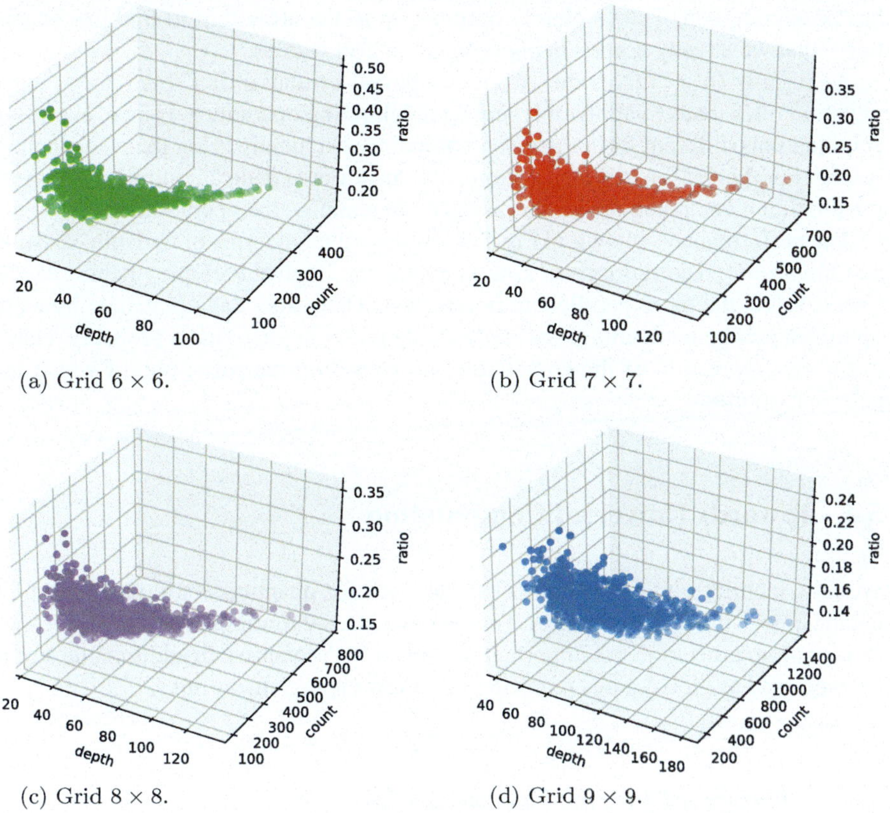

(a) Grid 6×6.

(b) Grid 7×7.

(c) Grid 8×8.

(d) Grid 9×9.

Fig. 5.3 1000 Clifford circuits have been generated uniformly at random for each of 4 grid topologies. The compilation performance is shown as the triplet $\mathfrak{E}$-count, $\mathfrak{E}$-depth and their ratio

that the grid lattice is a good family of lattices because of their ratio edges-to-nodes. Hence, we now consider 4 different grid lattices.

Any generated Clifford circuit is pre-processed to its normal form. As before, we tackle the compilation problem as the problem of finding minimum Steiner Trees that define how Global Gates are performed throughout the network. More specifically, since the $\wedge(\mathrm{Z})$ has no notion of control nor target, whenever two operators share a processor, we can compile the corresponding global gate by looking for a Steiner tree rooted in the common processor.

As the circuit made of $\wedge(\mathrm{Z})$ is fully commuting, we can select the operations according to a *greedy criterion*, which is also optimal. Namely, at each time-step the qubit involved in the highest number of operations is pivoting the iteration. All the qubits interacting with the pivot are gathered, together with the pivot, which is going to be the root of the minimal Steiner tree. This is repeated until each operation has been compiled. This is clearly optimal, up to the algorithm to find the Steiner tree—that we already argued being optimal on grid lattices.

As regards $\mathcal{C}_{\wedge(\mathtt{X})}$, this comes from an optimal synthesis. For this reason, commutation rules tends to apply only among neighboring operations. Hence, exploring combinations different from the given optimal would luckily bring to unsatisfactory results. For this reason, each operation $\wedge(\mathtt{X})$ has its own dedicated time-step, where the minimum entanglement path is assigned to compute the operation.[2]

The performance—shown in Fig. 5.3—is described in terms of $\mathfrak{E}$-depth, $\mathfrak{E}$-count and *their ratio*. The ratio gives a measure for the hardware employment. I.e., the lowest value, the more parallelism is achieved. The reader can appreciate how our optimizer gets better results as the hardware increases in size. This is probably related to the decomposition of the compilation problem in three parts, two of which are solved at optimum.

5.4 Towards Universal Computation

We now identify unitary operators of relevance for quantum computation and for entanglement-based architectures. There exist, in fact, unitaries that are hard to realize in a quantum processor, but important in application scenarios. In addition, these can be expressed as global gates requiring just a few entanglement links.

5.4.1 *Universal Circuit Decomposition*

We can achieve universality by using diagonal phase polynomials $\mathtt{D}_i$

$$\mathtt{D}_i\,|\mathtt{b}\rangle \equiv e^{\frac{i\pi}{4} f_i(\mathtt{b})}|\mathtt{b}\rangle, \tag{5.2}$$

linear reversible boolean functions $\mathtt{A}_i$

$$\mathtt{A}_i\,|\mathtt{b}\rangle \equiv |g_i(\mathtt{b})\rangle \tag{5.3}$$

and local gates. Specifically, any unitary $\mathtt{U}$ can be decomposed as in Fig. 5.4, where local operators are interleaved by unitaries $\mathtt{U}_i$, such that

$$\mathtt{U}_i\,|\mathtt{b}\rangle \equiv \mathtt{D}_i\mathtt{A}_i\,|\mathtt{b}\rangle \equiv e^{\frac{i\pi}{4} f_i(\mathtt{b})}|g_i(\mathtt{b})\rangle. \tag{5.4}$$

Starting from the decomposition given in Fig. 5.4, it is also possible to build a normal form which makes use of auxiliary qubits that inject Hadamard gates [54].

---

2 The depth could be optimized with the employment of quasi-parallelism. However, we are now considering topologies with capacity $c = 1$.

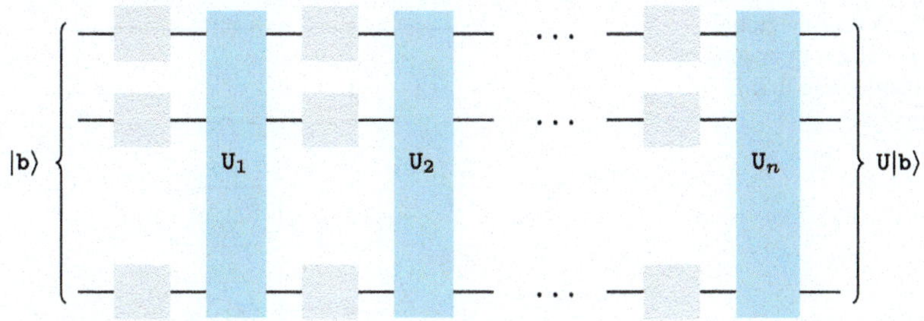

Fig. 5.4 Unitary decomposition for universal computation

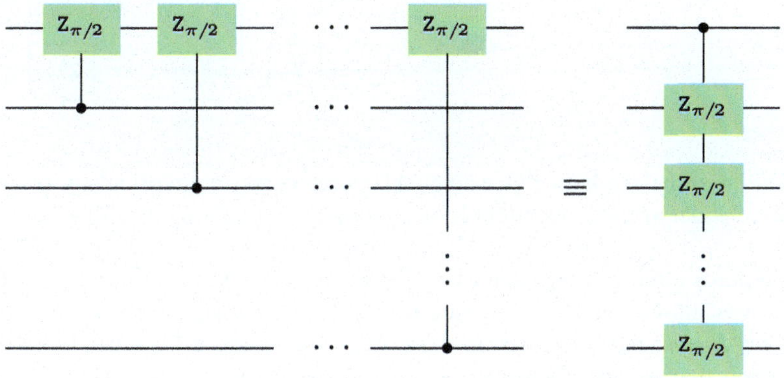

Fig. 5.5 A cascade of m two-qubit gates $\wedge(Z_{\pi/2})$ is equivalent to a single global gate $\wedge(Z_{\pi/2}^{\otimes m})$

The resulting circuit is a single occurrence of a unitary coming from the group of Equation (5.4), followed by a sequence of measurement-based Clifford operators. This justifies our interest into the optimization of unitaries from Eq. (5.4). Further motivations come from the fact that each of the subject operators—i.e. $\wedge(Z_{\pi/2}^{\otimes m})$, $\wedge(X_{\pi}^{\otimes m})$ and $\wedge_2(Z_{\pi}^{\otimes m})$.—is universal on its own, by adding the single-qubit unitary group $U(2)$ to the generator set.

5.4.2 *Circuit Synthesis and Optimization*

We now propose a method to realize unitaries from Eq. (5.4) with global gates.

First of all, we rule out from the modeling any local operator, as there is no need for link optimization. Our goal is therefore to build global gates for the following three global gates: $\wedge(Z_{\pi/2}^{\otimes m})$, $\wedge(X_{\pi}^{\otimes m})$ and $\wedge_2(Z_{\pi}^{\otimes m})$.

Starting from $\wedge(Z_{\pi/2})$ gates, as these gates are diagonal, it is trivial to construct global gates by re-arranging the circuit into one composed by operators in cascade—see Fig. 5.5. Specifically, we select as pivot qubit the one that has the higher number of

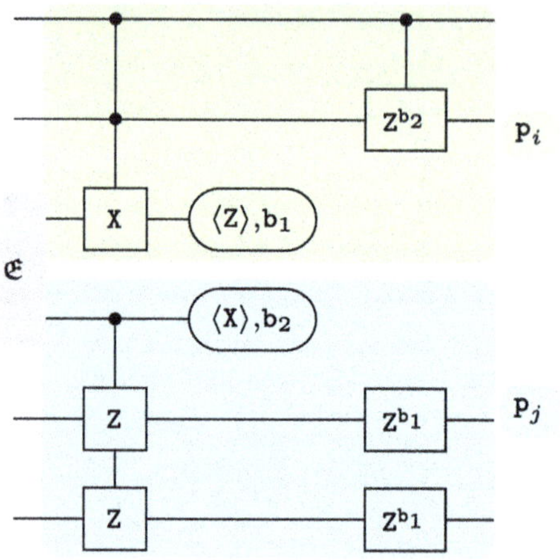

Fig. 5.6 A remote global gate $\wedge_2(Z \otimes Z)$, by means of one entanglement link [55]

interactions with the others. Since diagonal operators are symmetrical, the pivot qubit can be considered the control qubit of a global gate targeting to all the interacting qubits.

As regards circuits composed of $\wedge(X_\pi)$ gates, these do not commute in general. Hence, we cannot re-arrange the gates arbitrarily as for diagonal phase polynomials. What we do, instead, is gathering those gates that share the control qubit (or the target) and that are in proximity one another, within the circuit. This is a greedy selection.

The last kind of global gate is $\wedge_2(Z_\pi^{\otimes m})$. This is a bit trickier to generate, as there are several ways one can approach the problem. Similarly to what we do for $\wedge(Z_{\pi/2})$ gates, we select two pivots, which play the role of control qubits. We want the pivots to belong the same node—see Fig. 5.6. In case this does not occur, we teleport one or both qubits.

In conclusion, for any given global gate, we compute an entanglement tree by solving the minimum Steiner tree problem. As mentioned before, a global gate corresponding to 3-degree monomials may have the two control qubits allocated on different processors. In such a case, before computing an entanglement tree, we look for a shortest entanglement path that teleports the control qubits to the same core.

5.4.3 *Circuit Cost Analysis*

Here we report the results we obtained with our circuit synthesis and optimization strategies. A simulation run can be summarized in the following steps:

1. Sample generation (graphs and circuits).
2. Global gate synthesis.
3. Entanglement trees construction.

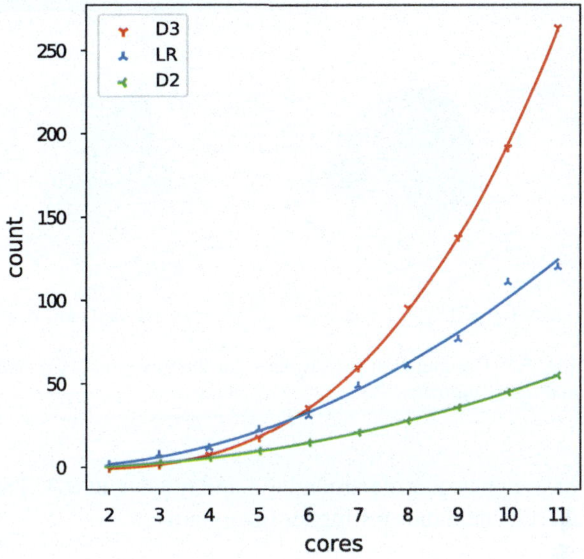

Fig. 5.7 Link count for each group: 3-degree monomials (red), 2-degree monomials (green) and linear reversible boolean functions (blue)

The outcome of a simulation is the *count* of links requested by the entanglement trees. The count is an index for the resource load required to run a circuit. Figure 5.7 shows the links count for the three different groups of circuits. When scaling the number of processors, more combinations are available and the circuit size increases. It is for this reason that the count for 3-degree monomials grows the most, followed by linear reversible boolean functions and, then, 2-degree monomials.

With the next round of experiments, we evaluate under what assumptions, entanglement-based architectures are in line with *noisy intermediate-scale quantum* paradigm [56]. Quantum noise has high impact on such architectures, impeding to achieve a fidelity arbitrarily high. In accordance, we estimate the *computation infidelity*, varying the quality of entanglement links—i.e. [0.995, 1.000]. In other words, we sweep the link fidelity.

Another axis represents the hardware, which we assume to be made of processors, arranged in grid disposition. Each processor has the minimal required resources, i.e., two computation qubits and four communication qubits.

For each unitary group and hardware size, we generate a sample of 10 circuits[3] and take the average count. Finally, for a given hardware size, link fidelity and count, we calculate the computation infidelity—see Fig. 5.8.

It can be observed that 2-degrees monomials—Fig. 5.8b—have good performance, with an infidelity in average < 0.1 and, for the 75% of the population—i.e. the *upper quartile*—, the infidelity is < 0.2. A bit less performing are the linear reversible boolean functions—Fig. 5.8c—, with median below 0.2 and upper quartile roughly at 0.4. The most concerning results regard 3-degree monomials—Fig. 5.8a—, which

[3] Which we verified being big enough, as the standard deviation is small.

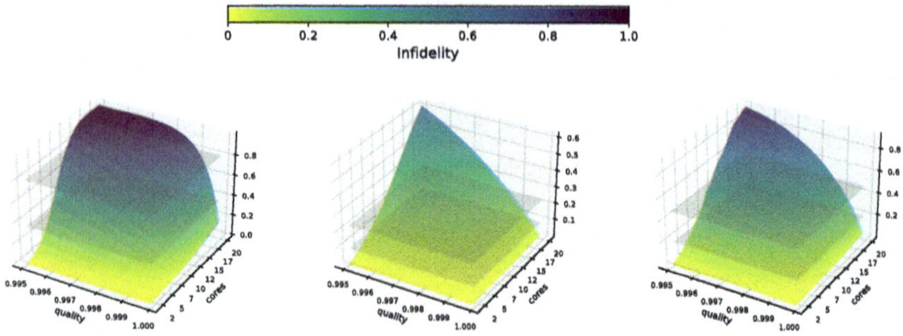

(a) 3-degree diagonal monomials. (b) 2-degree diagonal monomials. (c) Reversible boolean functions.

Fig. 5.8 Computation performance for the three unitary groups. The hyperplanes denote *median* and *upper quartile*

have upper quartile around 0.8 and median at 0.3. This means that computing such circuits demands for higher link quality.

5.5 Conclusion: The Importance of Layering the Architecture

According to our envision of the full-stack development—see Sect. 1.2—compiler and scheduler take care of creating a logical circuit which is compliant with the hardware. To understand how helpful a software layer is, it is important to keep in mind that its role should be interpreted in two different ways.

1. Lower layers guarantee a reliable abstraction on which one can build a logical paradigm.
2. Upper layers lighten the lower layers from all the tasks which are merely logical.

Specifically to the second interpretation, consider that each layer of the stack has important and complex tasks to care about. Lower layers are more prone to treat real-time problems. In fact, the scheduler cares about synchronization and connectivity maximization. The hardware layer cares of being efficient in terms of, e.g., pushing the technologies at their maximal performance to get high fidelities and high success rates. The above tasks are for sure hard. For this reason, identify what can be **delegated** to a logical reasoner—i.e. compiler and scheduler—may bring critical advantages to the overall architecture.

To get a practical intuition of what we are stating, consider for example our assumption for the $\mathcal{E}$ operator. We used this assumption to provide the reader with a model relatively easy to understand. But with a careful knowledge of the underlying architecture from an information theory perspective, something better can be achieved. As an example, consider the stationary-flying system—see Chap. 1—generating and distributing entangled states. When it succeeds, it generally means that a heralded Bell state has been produced—i.e. $\{|\Psi^+\rangle, |\Psi^-\rangle, |\Phi^+\rangle, |\Phi^-\rangle\}$.

Now; if the reader believe, as we do, that delegating to the compiler to analyse post-processing will bring advantages to the general performance of the quantum computation, then it is clear that this approach does not stop to the models we created throughout this chapter. As a matter of fact, Bell states differ one another by Pauli corrections, which are usually treated at the hardware layer to provide the upper layers with $|\Phi^+\rangle$. But, even if local operations are very efficient, when it comes to multiple repeated steps—as for the case of quantum computation—every single gate avoided has a positive impact on the final fidelity. For the case we are considering now, the presence of a compiler enable the hardware to delegate the corrections, which now includes them to the big set of logical instructions to optimize. As basic example consider the circuit in Fig. 5.9, where a $\wedge(\mathtt{X})$ runs with a different Bell state.

Let us see numerically what happens without delegating the correction to the compiler. Consider a single-qubit gate error of probability $\mathtt{p} = 10^{-n}$. Then, for $10 \cdot m$ non-local operations, the probability get worse of m orders of magnitude, i.e. 10^{-n+m}. On contrary, the compiler eliminates all the corrections, preserving the error probability to 10^{-n}. The same reasoning applies to the running-time.

With our models we covered several interesting group of circuits, for which the compiler eliminates every single correction from the quantum computation. The Bell state correction is no different. Specifically, instead of performing Pauli-corrections, the compiler keeps track of the logical error propagation caused by avoiding it. At the end of the computation the compiler provides the classical computer the necessary bit-flip corrections to get the right final state.

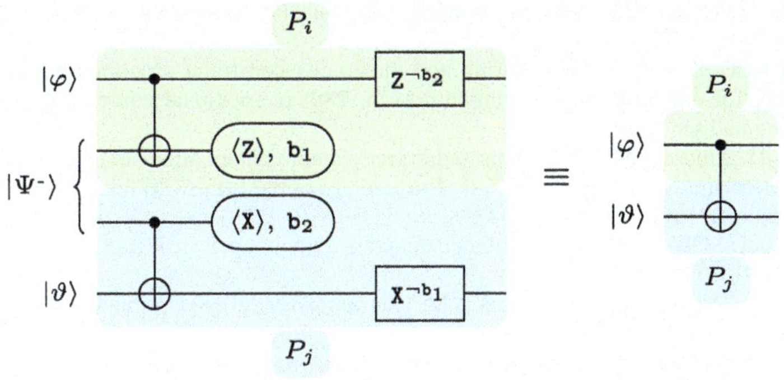

Fig. 5.9 Non-local $\wedge(\mathtt{X})$ performed by means of $|\Psi^-\rangle$. This example shows how to avoid Pauli corrections

References

1. Botea A, Kishimoto A, Marinescu R (2018) On the complexity of quantum circuit compilation. In: Eleventh annual symposium on combinatorial search
2. Madden L, Simonetto A (2022) blackBest approximate quantum compiling problems. ACM Trans Quantum Comput 3(2):1–29
3. Stefan H, Alwin Z, Robert W (2021) Exploiting quantum teleportation in quantum circuit mapping. In: 26th Asia and South Pacific design automation conference (ASP-DAC). IEEE, pp 792–797
4. Lukas B, Sarah S, Robert W (2022) Limiting the search space in optimal quantum circuit mapping. In: 27th Asia and South Pacific design automation conference (ASP-DAC). IEEE, pp 466–471
5. Maslov D, Falconer SM, Mosca M (2008) Quantum circuit placement. IEEE Trans Comput-Aided Design Integrat Circuits Syst 27(4):752–763
6. Siraichi MY et al (2018) Qubit allocation. In: Proceedings of the 2018 international symposium on code generation and optimization, pp 113–125
7. Wille R, Burgholzer L, Zulehner A (2019) Mapping quantum circuits to IBM QX architectures using the minimal number of `SWAP` and `H` operations. In: 2019 56th ACM/IEEE design automation conference. IEEE, pp 1–6
8. Li G, Ding Y, Xie Y (2019) Tackling the qubit mapping problem for NISQ-era quantum devices. In: Proceedings of the 24th international conference on architectural support for programming languages and operating systems, pp 1001–1014
9. Zulehner A, Wille R (2019) Compiling $SU(4)$ quantum circuits to IBM QX architectures. In: Proceedings of the 24th Asia and South Pacific design automation conference, pp 185–190
10. Itoko T et al (2019) Quantum circuit compilers using gate commutation rules. In: Proceedings of the 24th Asia and South Pacific design automation conference, pp 191–196
11. Zhang Y-H et al (2020) Topological quantum compiling with reinforcement learning. Phys Rev Lett 125(17):170501
12. Karalekas PJ et al (2020) A quantum-classical cloud platform optimized for variational hybrid algorithms. Quantum Sci Technol 5(2):024003
13. Moro L et al (2021) Quantum compiling by deep reinforcement learning. Nat Commun Phys 4(178)
14. Maronese M et al (2022) Quantum compiling. In: Quantum computing environments. Springer, pp 39–74
15. Booth KEC et al (2018) Comparing and integrating constraint programming and temporal planning for quantum circuit compilation. In: 28th international conference on automated planning and scheduling
16. Ferrari D, Amoretti M (2022) Noise-adaptive quantum compilation strategies evaluated with application-motivated benchmarks. In: Proceedings of the 19th ACM international conference on computing frontiers, pp 237–243
17. Beals R et al (2013) Efficient distributed quantum computing. Proc R Soc A: Math, Phys Engin Sci 469(2153):20120686
18. Zomorodi-Moghadam M, Houshmand M, Houshmand M (2018) Optimizing teleportation cost in distributed quantum circuits. Int J Theoret Phys 57(3):848–861
19. Daei O, Navi K, Zomorodi-Moghadam M (2020) Optimized quantum circuit partitioning. Int J Theoret Phys 59(12):3804–3820
20. Davide F et al (2021) Compiler design for distributed quantum computing. IEEE Trans Quantum Engin 2:1–20
21. Nikahd E et al (2021) Automated window-based partitioning of quantum circuits. Phys Scrip 96(3):035102
22. Dadkhah D, Zomorodi M, Hosseini SE (2021) A new approach for optimization of distributed quantum circuits. Int J Theoret Phys 60(9):3271–3285
23. Daei O, Navi K, Zomorodi M (2021) Improving the teleportation cost in distributed quantum circuits based on commuting of gates. Int J Theoret Phys 60(9):3494–3513

24. Sarvaghad-Moghaddam M, Zomorodi M (2021) A general protocol for distributed quantum gates. Quantum Inf Process 20(8):1–14
25. Zomorodi-Moghadam M, Davarzani Z, Ghodsollahee I et al (2021) Connectivity matrix model of quantum circuits and its application to distributed quantum circuit optimization. Quantum Inf Process 20
26. Sundaram RG, Gupta H, Ramakrishnan CR (2021) Efficient distribution of quantum circuits. In: 35th international symposium on distributed computing. Schloss Dagstuhl-Leibniz-Zentrum für Informatik
27. Stephenson LJ et al (2020) blackHigh-rate, high-fidelity entanglement of qubits across an elementary quantum network. Phys Rev Lett 124(11):110501
28. Cuomo D, Caleffi M, Cacciapuoti AS (2020) Towards a distributed quantum computing ecosystem.IET Quantum Commun 1(1):3–8
29. Van Meter R, Devitt SJ (2016) The path to scalable distributed quantum computing. Computer 49(9):31–42
30. Kleinberg J, Tardos E (2006) Algorithm design. Pearson Education, India
31. Cieslik D (2013) Steiner minimal trees, vol 23. Springer Science & Business Media
32. Kou L, Markowsky G, Berman L (1981) A fast algorithm for Steiner trees. Acta Inf 15(2):141–145
33. Berman P, Karpinski M, Zelikovsky A (2009) 1.25-approximation algorithm for steiner tree problem with distances 1 and 2. In: Workshop on algorithms and data structures. Springer, pp 86–97
34. Chlebík M, Chlebíková J (2008) The Steiner tree problem on graphs: inapproximability results. Theoret Comput Sci 406(3):207–214
35. Byrka J et al (2010) An improved LP-based approximation for Steiner tree. In: Proceedings of the forty-second ACM symposium on Theory of computing, pp 583–592
36. Brazil M et al (1996) Minimal Steiner trees for 2k× 2ksquare lattices. J Combinat Theory, Ser A 73(1):91–110
37. Brazil M et al (1997) Minimal Steiner trees for rectangular arrays of lattice points. J Combinat Theory, Ser A 79(2):181–208
38. Gottesman D (1998) Theory of fault-tolerant quantum computation. Phys Rev A 57(1):127
39. Selinger P (2013) Quantum circuits of texttttT-depth one. Phys Rev A 87(4):042302
40. Amy M, Maslov D, Mosca M (2014) Polynomial-time `T`-depth optimization of Clifford+`T`circuits via matroid partitioning. IEEE Trans Comput-Aided Design Integrat Circuits Syst 33(10):1476–1489. matroid partitioning. IEEE Trans Comput-Aided Design Integrat Circuits Syst 33(10):1476–1489 (2014)
41. Hagberg AA, Schult DA, Swart PJ (2008) Exploring network structure, dynamics, and function using `networkx`. In: Proceedings of the 7th python in science conference. Pasadena, CA USA, 2008, pp 11–15
42. Duncan R et al (2020) Graph-theoretic simplification of quantum circuits with the `ZX`-calculus. Quantum 4:279
43. Kissinger A, van de Wetering J (2020) PyZX: large scale automated diagrammatic reasoning. In: Proceedings 16th international conference on quantum physics and logic, vol 318. Open Publishing Association, pp 229–241
44. van de Wetering J (2020) `ZX`-calculus for the working quantum computer scientist. arXiv preprint arXiv:2012.13966
45. Coecke B, Duncan R (2007) A graphical calculus for quantum observables. https://zxcalculus.com/publications.html
46. Cuomo D et al (2023) Optimized compiler for distributed quantum computing. ACM Trans Quantum Comput
47. De Brugiere TG et al (2021) Reducing the depth of linear reversible quantum circuits. IEEE Trans Quantum Engin 2:1–22
48. Bravyi S, Maslov D, Nam Y (2022) Constant-cost implementations of clifford operations and multiply controlled gates using global interactions. arXiv preprint arXiv:2207.08691
49. Lu Y et al (2019) Global entangling gates on arbitrary ion qubits. Nature 572(7769):363–367

50. Casanova J et al (2012) Quantum simulation of interacting fermion lattice models in trapped ions. Phys Rev Lett 108(19):190502
51. Ivanov SS, Ivanov PA, Vitanov NV (2015) Efficient construction of three-and four-qubit quantum gates by global entangling gates. Phys Rev A 91(3):032311
52. Martinez EA et al (2016) Compiling quantum algorithms for architectures with multi-qubit gates. New J Phys 18(6):063029
53. van de Wetering J (2021) Constructing quantum circuits with global gates. New J Phys 23(4):043015
54. Heyfron LE, Campbell ET (2018) An efficient quantum compiler that reduces Tcount. Quantum Sci Technol 4(1):015004
55. Sarvaghad-Moghaddam M, Zomorodi M (2021) A general protocol for distributed quantum gates. Quantum Inf Process 20(8):265
56. Preskill J (2018) Quantum computing in the NISQ era and beyond. Quantum 2(79)

The manufacturer's authorised representative in the EU is Springer Nature Customer Service Centre GmbH, Europaplatz 3, 69115 Heidelberg, Germany. If you have any concerns regarding our products, please contact ProductSafety@springernature.com

Printed and bound by CPI Group (UK) Ltd, Croydon, CR0 4YY

30/09/2026

02225471-0002